Fundamentals and Applications of Hardcopy Communication

Joceli Mayer · Paulo V. K. Borges
Steven J. Simske

Fundamentals and Applications of Hardcopy Communication

Conveying Side Information by Printed Media

 Springer

Joceli Mayer
Department of Electrical Engineering
Universidade Federal de Santa Catarina
Florianopolis, Santa Catarina, Brazil

Steven J. Simske
Colorado State University
Fort Collins, Colorado, USA

Paulo V. K. Borges
Robotics and Autonomous Systems
Group, CSIRO
Canberra, ACT, Australia

QCAT - Queensland Centre
for Advanced Technologies
Pullenvale, QLD, Australia

ISBN 978-3-030-08915-3   ISBN 978-3-319-74083-6   (eBook)
https://doi.org/10.1007/978-3-319-74083-6

We dedicate this book to our families.

Acknowledgments

We are indebted to a number of individuals in academic circle as well as in industry and government who have contributed to the preparation of this book. We wish to extend our appreciation for the financial support provided by Brazilian R&D funding government agencies (CNPq, CAPES, FINEP) that enable us to participate in scientific conferences and provided scholarships, equipments, and materials. Many thanks to our private partners (Hewlett-Packard Company, Intelbras S.A.) and the challenging projects from which we investigate interesting problems, and some findings are presented in this book. We are also thankful to Data61 in the CSIRO for the financial support. We would like to acknowledge the support and motivation provided by the Springer Team, especially from Jorge Nakahara Jr., Springer computer science editor.

Contents

Acronyms

A/D	Analog to digital
AWGN	Additive white Gaussian noise
BCH	Bose, Chaudhuri, and Hocquenghem
BER	Bit error rate
BW	Black and white
CAD	Computer-aided design
CCD	Charge-coupled device
CDM	Code division modulation
D/A	Digital to analog
DFT	Discrete Fourier transform
ECC	Error correction code
EM	Expectation-maximization
FFT	Fast Fourier transform
GUI	Graphic user interface
HCCB	Microsoft High Capacity Color Barcode
HSV	Hue, Saturation, Value Color Model
HVS	Human vision system
IPS	Inkjet Print Scan
IQM	Index quantization modulation
LOG	Laplacian of Gaussian filter
NP	Neyman-Pearson
NPI	Non-payload indicia
OCR	Optical character recognition
PAM	Pulse amplitude modulation
PBC	Position-based coding
PDF	Probability density function
PFN	Probability of false negative
PFP	Probability of false positive
PRS	Pseudo-random sequence
PS	Print and scan
PSD	Power spectral density

PSNR	Peak signal-to-noise ratio
PSWR	Peak signal-to-watermark ratio
RS	Reed-Solomon
RSA	Ron Rivest, Adi Shamir, and Leonard Adleman algorithm
SM	Sequential modulation
SSIM	Structural Similarity Index Metric
TCM	Text color modulation
TDM	Time division modulation
THM	Text halftoning modulation
TLM	Text luminance modulation
TSA	Text self authentication
VDP	Variable data printing
XOR	Exclusive OR

Chapter 1
Introduction

This book discusses techniques to convey side information over a printed media. Those techniques will be referred to as hardcopy communication techniques. The information is modulated over the digital host document and then printed using laser, inkjet printer technologies or other printing/branding processes. In order to recover the information at receiver side, the printed document or branded product is digitalized using either cameras or desk scanners and the resulting digital image is processed to decode the embedded information. The embedded information can be exploited for a variety of applications: document, user and/or content authentication, copyright protection, piracy deterrent, document tracking and other uses.

This book aims to provide a solid understanding of the issues of hardcopy communication and the novel applications from this recent technology, still on the making, by examining techniques, deriving analyses, providing examples for some applications. The related research on digital watermarking provides a mature technology [1, 3–10] from which many techniques can be also exploited in hardcopy communication. These approaches used in digital watermarking are usually designed to survive one or many so called attacks, namely, additive noise, rotation, geometric distortions and others. This mature technology provides a large and solid literature on how to deal with ill-intentioned attackers who desire to remove, substitute, decode, without permission or a secret key, the hidden information embedded on the host document. Mostly watermarking techniques presented in literature focus mainly on digital data such as speech, image, video or digital documents.

A greater challenge arises when analogue world is posed between the digital side information embedder and the digital decoder. Thus, our task is to describe effective techniques that enable hardcopy communication over printed paper. These techniques are designed to address the sources of distortion on the channel, namely the print and scan (PS) channel that includes all sorts of effects. These effects are originated from the mechanical imperfections of the printer, optical system,

© Springer International Publishing AG 2018
J. Mayer et al., *Fundamentals and Applications of Hardcopy Communication*,
https://doi.org/10.1007/978-3-319-74083-6_1

properties of laser and inkjet technologies, aging and user handling of the document, paper media and ink/toner chemical properties and interaction and others sources that need to be understood to achieve some level of hardcopy communication.

Some people may consider hardcopy communication to convey side information as a special case of digital watermarking technology, yet the hardcopy communication techniques differ from the digital problem due to the analogue channel and the sources of distortions. To achieve hardcopy communication one needs to design techniques that enables the side information to survive the channel (be able to decode at receiver side) considering the payload required for the application and usually also to survive malicious attackers, the main concern of digital watermarking.

Researchers working to achieve effective hardcopy communication find quite challenging issues but also newer and interesting applications for real world problems. It is a fascinating task to pursue as the world is moving from local services to distributed services in analogue domain, we find that hardcopy communication plays a important role in the future of the so called "internet of things" technology. The embedded information into paper enables distributed services as tracking of products distributed worldwide, even food, by automatic visual inspection with cameras and/or scanners. Moreover, plenty of applications arise from hardcopy communication, for example:

- Enforcing copyright of printed media by inserting overt or covert marks that survive the attacks and the print and scan channel along with the external distortions (user handling, spills, aging, etc.).
- Piracy deterrent by inserting a smart label (either printed on a piece of paper or on the product material) for computer chips, foods, pharmaceutical drugs, high value sunglasses and so on.
- Content authentication by validating information in documents to avoid frauds of many types.

These examples are a few of many possible applications for the hardcopy communication technology. Notice that properly designing solutions for the afore-mentioned applications have a huge positive impact on the economy of many leading industries such as the electronic companies, pharmaceutical laboratories, food industry, high value cosmetic or fashion brand products industry, and many others.

The need for hidden information has never been more prevalent. Customization and personalization are two related and extremely important trends in both 2D (traditional) and 3D (additive manufacturing) printing. In 2D printing, variable data printing (VDP), powered by the massive processing and data capabilities of modern digital presses, allows every label, every package, every document, and every piece of marketing collateral to contain personalized information. This information can include multiple serialized VDP marks, including barcodes, digital watermarks and graphical alphanumerics. Importantly, each of these marks can be assigned to a different workflow; and by combining with a secure log-on, [each can be] associated with different services and allowed actions based on the role a person plays in the workflow. For example, a barcode can be used for point of sale (cashier), inventory control (stock person), track and trace (warehouser, distributor), and couponing/gaming/loyalty points (customer). However, if the forensics of the barcode are captured, the same barcode can also be used for determining counterfeits (by a secret shopper) or for determining smuggling and other forms of fraud (investigatory agent). The point is, multiple sources of variability allow various a priori and a posteriori applications to be defined. Hidden information is a crucial part of this variability, since hidden information can be used where aesthetics are important.

In 3D printing and other forms of additive manufacturing, a whole new world of fraud possibilities are opened up. Text, images and music have long been the target of fraudulent agents, using everything from copying machines to Napster to Bit Torrent. When the design of hardware and other 3D objects *from circuits to sculpture* becomes as simple as loading the CAD drawings into a hybrid 3D printer, there will be previously non-existent considerations for liability, copyright and fair use. Having the ability to overtly, semi-covertly and covertly mark each and every custom manufactured or personalized manufactured part will be an important way to address these three issues. Liability will be assessable when parts fail by authenticating the source of all parts in a system, whether they directly failed, translated failure to adjacent parts (e.g. through stress shielding), or can be exonerated from blame for failure. Copyright can be determined by reading information from the part and associating it with *open source* hardware or branded hardware. Fair use, finally, can be assessed by considering the relative mixture of branded, open source and locally manufactured parts.

In the near future, every object printed or otherwise additively manufactured is going to have information associated with it in the construction process. Being able to encode information onto/into these objects for later reliable and secure decoding is an important part of the future of printing and manufacturing. The techniques in this book help to address this future.

This book provides an overview of three main approaches to convey information over printed media, namely, image color watermarking, text modulation and color barcodes. Examples, analyses and a variety of techniques for encoding, segmentation, code correction, detection and others are also discussed in details. The book focuses on hardcopy communication by presenting statistical modeling, encoding and decoding, and other techniques properly tuned for the print and scan channel. Our goal is to contribute to researchers by compiling a sound material on hardcopy communication that helps to build a solid basis on this field of communication which enables a variety of very interesting and high value applications.

The intended audience for this book includes researchers and practitioners on the topic as well as graduate students looking for a guideline material to investigate the topic of Hardcopy Communication.

1.1 Overview

Chapter 2 discusses hardcopy communication by using printed images (either BW or color) as the host. Models and analyses of print and scan channels are discussed along with fundamentals and techniques to address these challenging channels for the purpose of communicating over printed images. Review on the state-of-the-art and approaches are given to illustrate to researchers the issues, the techniques and the expected performance from communication using images over printed paper. Covert techniques for communication are discussed by two examples, one is aimed to address multiple black and white print copies (scanning followed by printing) that may arise in regular use of printed documents in offices. The goal is to be able to decode the side information using a minimal amount of embedding energy, that may be seen as distortion of the host document. Theoretical models and assumptions are described in details and experimental results are provided. The second example is aimed to convey information in printed images through color inkjet print channels by designing special patterns, employing a color rejection approach and exploiting the concept of informed coding inspired by the work of Prof. Max H. Costa, namely, the "Writing in Dirty Paper" [2]. Analysis and results are provided to illustrate the effectiveness of the techniques discussed.

Chapter 3 discusses communication over printed paper by using text modulation. Conveying information hidden in text documents is very important for many applications, including content authentication to fraud determent, information hiding for various purposes such as copyright and tracking, etc. Many techniques from the literature on text modulation are discussed including halftoning, affine transformations and text luminance. Advanced techniques for text luminance approach are discussed along with a practical authentication protocol for encryption and decryption considering noise to model distortions in the printed media and other sources. A position based approach for text modulation is presented and the use of color for text modulation is also evaluated. The chapter provides a variety of techniques for text modulation to convey side information over paper, performance analyses and practical examples are also presented.

Chapter 4 discusses overt techniques as an alternative to convey information. Visible color barcodes specially designed to survive the PS color channel are employed to convey information. Overt techniques (in contrast to covert techniques discussed so far) usually are not meant to be transparent as the main goal is to achieve efficient high payload communication (smallest print code in the document) while surviving the print and scan channel along with external distortions. Literature review and special techniques needed for these codes are discussed along with examples of typical distortions due to accidental pen scribing, spills and smudging. An example of decoding technique is presented. It is based on the expectation-maximization (EM) algorithm by employing a statistical model for the printed colored dots. In order to address external distortions imposed by human handling of the paper media, error correction codes are designed to help the embedded information survive the aforementioned distortions and be decoded by the receiver.

Tailored segmentation techniques along with proposed visual cues are also required to decode the information. Results are provided to illustrate the capacity of these techniques to convey information over color print codes and the additional performance achieved by using error correction codes.

References

1. M. Barni, F. Bartolini, *Watermarking Systems Engineering* (Marcel Dekker, New York, 2004)
2. M.H.M. Costa, Writing on dirty paper. IEEE Trans. Inf. Theory **IT-29**, 439–441 (1983)
3. I.K. Cox, M.L. Miller, J.A. Bloom, *Digital Watermarking* (Morgan Kaufmann, San Francisco, 2002)
4. I.K. Cox, M.L. Miller, J.A. Bloom, J. Fridrich, T. Kalker, *Digital Watermarking and Steganography*, 2nd edn. (Morgan Kaufmann, San Francisco, 2008)
5. B. Furht, D. Kirovski, *Multimedia Watermarking Techniques and Applications* (Auerbach Publications, Boston, 2006)
6. B. Furht, E. Muharemagic, D. Socek, *Multimedia Encryption and Watermarking* (Springer, New York, 2005)
7. J. Seitz (ed.), *Digital Watermarking for Digital Media* (Information Science Publishing, Hershey, 2005)
8. F.Y. Shih, *Digital Watermarking and Steganography* (CRC Press, Boca Raton, 2008)
9. F.-H. Wang, *Innovations in Digital Watermarking Techniques* (Springer, Berlin, 2009)
10. P. Wayner, *Disappearing Cryptography* (Morgan Kaufmann, San Francisco, 2009)

Chapter 2
Hardcopy Image Communication

The literature presents an extensive research on digital watermarking applications with digital image, speech, music and video media. However, much less research effort has been applied to hardcopy communication via printed media. Hardcopy images can be used to convey side information required for many security, tracking, copyright and forensic applications. Many office applications rely on a tight connection between physical and electronic documents. Physical (printed) images can be a token for the electronic document (or documents) it represents directly (or indirectly as part of a shared workflow). Hiding information in images to be printed can be used to embed side information or to save real estate on the label, document or packaging printed media. This hidden information is useful for location-based services, point-of-sale, security, counterfeit and piracy deterrence, content authentication, fingerprinting and more. Depending on the application different approaches to address encoding, modulation, and detection of information are required to achieve a given tradeoff among payload, transparency and robustness. This chapter discusses state-of-the-art approaches for hardcopy image communication, describes main sources of distortion in the black-and-white and color print-scan channels, discusses resulting performance of alternative encoding, modulation and detection techniques.

2.1 Introduction

Transmission of information over the printed media is quite challenging [23] due to the various distortions existing in the print-scan channel. This channel introduces distortions into the black-and-white or color patterns transmitted along with the host image to convey some side information. Some of the challenging distortions are those originated from ink property variations such as spreading and mixing, from variations in the coated media properties, from the optical and mechanical disturbances at printing and scanning devices [3] and from scanning sensor responses. Moreover, as the printed media may be subjected to external distortions as scraping, humidity, finger smudging and aging of the paper media and the ink/toner, we need robust and advanced techniques to address these distortions as the techniques designed for digital media alone cannot deal with them.

2.1.1 A Communication Problem

The task of conveying information by printed media, also known as hardcopy communication, can be understood as a communication problem where the host image is modulated by some means to convey information over a print-and-scan channel, which later on is decoded possibly with the help of some post-processing techniques. Moreover, the choice of techniques for coding and embedding an information into the host image and decoding of the transmitted information will depend on the specific application and on the characteristics of the channel involved. The inkjet printer distortions differs considerably from the laser printer effects, and usually very different techniques are required for each type of branding device. The pixel resolution, optical lens set quality and mechanical precision of the scanning device also affect considerably the robustness of the transmitted information. Moreover, when the scanning device available is a camera, standalone or in a cell phone, a very different problem is set with a different tradeoff between robustness and payload, requiring enhanced techniques for illumination and color equalization, usually employing geometrical synchronization with visual cues.

2.1.2 Information Encoding

The message bits, m, to be conveyed may represent a tracking number related to location, brand, user, to another content or to any additional information to be transmitted along with the image to the user. It may also represent some checksum number to authenticate some information in the image. For these cases, the information can be properly encoded into m' to achieve a more compact size (compression), to be hidden using a cryptography protocol, to add robustness using Error Correction Codes (ECC) to reduce the effects of channel errors in the bit error rate or to exploit an informed encoding approach as in [6]. In the latter case, the message may be encoded depending on an estimation of the channel robustness on the transmitted messages, providing a metric of the most robust encoding for that encoded message, m' and the host, $r(m', x)$. Thus, the informed coding approaches aim to select the best encoding for a message m, from a set of alternative and equivalent codes, M', that provides the best robustness over the channel perturbations. This set of equivalent codes results in the same message $m = d(m' \in M)$, where $d(\cdot)$ is the decoding function.

2.1.3   Embedding into the Host

The encoded information m' can be embedded (or modulated) into the host by using many approaches: (1) by changing the pixels amplitudes in the spatial domain of the host image; (2) by modifying coefficients of a transformed domain using Fourier, wavelets, Discrete Cosine or other transforms; (3) by disturbing the quantization levels of the pixel amplitudes employing an Index Quantization Modulation (IQM) approach; (4) by altering the least significant bits of the pixels; (5) by altering the color channels either in space or transformed domain; among other options. The resulting embedded image is $x' = f(x, m')$, where $f(x, m')$ is the embedding function that may depend only on the encoded message or also on the host image (informed embedding). The embedding function may explore some local information about the image either in the sample or frequency domain aiming either higher perceptual transparency and/or a desired payload. The choice of the embedding function depends on the requirements of the application which may enforce either a higher payload or a more transparent embedding. In general, the choice or design of an embedding function depends on the tradeoff among payload, robustness and transparency, which is defined by the application.

2.1.4   Decoding the Information

The decoding process may require some post-processing or equalization to remove host interference, to recover synchronism due to rotation, for instance, or to compensate for some linear or non-linear channel distortions before the actual process of decoding. The decoder architecture depends on the type of the modulation/encoding applied: additive spread spectrum modulation, quantization index modulation or other form of embedding/modulation that may use an error correction code and/or cryptography protocol. Thus, the decoder may require a secret key to decode message from the extracted string of bits.

In the following we provide a review on the literature on hardcopy watermarking and then describe two approaches for hardcopy communication to illustrate the techniques, performance and issues of communication over print and scan channels.

2.2   Review on Techniques for Hardcopy Image Communication

Designing a robust decoder to the conveyed side information over color printed media can be challenging [5, 10, 14, 22] due to various non-linear distortions present in a color print-scan channel. In particular, strong distortions may be originated by ink spreading and mixing existing in inkjet printers and other disturbances from the coated media properties, optical, mechanical and scanning sensor responses [3].

The literature provides various techniques to effective hardcopy communication over print-scan channels. The performance varies depending on the selected techniques for modulation, synchronization and detection, and also depends on the applications being targeted. Many approaches embed information only in the luminance channel or either exploiting or generating the printer halftoning. Other techniques embed into all color channels. Aiming to improve the performance of the decoder, specially designed approaches are employed to recover synchronism due to geometrical rotation originated in the scanning (digitalization) process. One approach is to introduce visual cues or synchronization blocks along with the embedded information into the marked image, either in the frequency domain or space domain through oriented patterns generated by the halftoning process in laser printers or even by locating the image borders with the help of the Hough transform [8]. Additionally to a proper modulation and synchronization techniques, an error correction code approach is usually employed at the message embedding in order to deal with the strong errors originated from different sources (spilling, aging, color spreading, etc.) in the analog channel.

A variety of approaches are available in the literature. For instance, the technique described in [12] conveys information by modulating the angle of oriented periodical sequences which are embedded into image spatial blocks. The approach requires one block to embed synchronization codes. This approach exploits only the luminance channel and does not consider either informed coding (to be describe latter) or exploit the color channel. Nevertheless, it provides a small payload of 40 bits per printed page.

Another approach provided in [22] embeds information into the luminance image phase spectrum using differential quantization index modulation. The printer halftoning is exploited to estimate the rotation. It achieves a bigger payload of hundreds of bits considering monochromatic images.

The work in [10] exploits adaptive block embedding into the Discrete Fourier Transform (DFT) magnitude blocks. A block is classified as either smooth or texture type. Next, a proper embedding technique is applied to each block type. The approach requires to detect the printed image boundaries for mark synchronization and it is achieved by using the properties of the Hough transform. As a result, the approach offer robustness to distortions in the print-scan channel and also to rotation, achieving a payload of over 1000 bits. However, the results indicate a high bit error rate (BER) over 10% when dealing with gray scaled images.

A different technique in [20] embeds a circular template into the Fourier transform magnitude domain to address the rotation and scaling resulting distortions due to the printing followed by scanning process. This work also embeds a special template in spatial domain to address translations but the message is modulated and then embedded in the wavelet domain. The resulting payload for this technique is over 100 bits by employing error correction algorithm. The resulting BER achieved is less than 2%. This technique, along with other is specially designed for gray scale images only.

Since the aforementioned techniques exploit and embed information only in the luminance channel, they do not take advantage of the color channels to increase robustness and payload as well as do not address the resulting distortions of embedding into these channels. Different approaches are required to exploit the color channels and address the resulting distortions due to analog color print scan channel.

For this purpose, the work in [9], embeds information into Discrete Fourier Transform only in the color red component. Another technique in [19] exploits informed halftoning modulation in the frequency domain. Although halftoning modulation can achieve high payload it also requires to control the printer driver in order to bypass the printer processing and halftoning, which calls for a new printer driver. This task of generating a new driver is quite cumbersome and also requires to distribute the resulting printer driver/firmware to the end users. Moreover, those techniques that exploit color channels for hardcopy communication by embedding into frequency domain [9] or exploiting halftoning [19] are specially designed only for laser print-and-scan channels. As a result, they are not efficient for other branding technologies as inkjet printing due to distortions such as the ink spreading and color mixing. These distortions smooth out low frequency marks as well as the embedded halftoning patterns.

In order to appreciate some of main issues when dealing with communication over print channels, it follows an approach designed to be robust to the print and scan channel even when it passes through the channel more than once.

2.3   Robust Authentication of Documents After Multiple Passes

This section addresses the design of a robust method for authentication of contents in a printed media which may have passed multiple times over the print and scan channel. The approach embeds the authentication message into the digital document before printing. The goal is to retrieve the message after receiving the printed and scanned work. This proposed method is aimed to be employed for all types of documents since the designed additive embedding approach is independent and robust to graphics, text or image components in the document. The approach improves the detection by employing segmentation of those interfering components (text, equations, graphics, drawings) that may exist in general printed documents. The remaining segmentation noise need to be addressed as it affects the detection performance as verified via analysis. Thus, an improved segmentation, which takes into account the size of the interfering connected components, is applied to mitigate this remaining noise. We employ M-ary modulation to encode the message and additively spread the resulting patterns over the digital document aiming to achieve low perceptual distortion. We also employ Time Division Modulation (TDM) to achieve a high robustness performance while embedding messages of sizes over 100 bytes per printed page. The resulting approach is robust to multiple black-and-white photocopies (multiple passes over the print-scan channel) of the document, as illustrated in the results. To achieve this performance the detector is adjusted accordingly to the print dot gain (dot spreading due to multiple scanning and printing). Improved results are achieved in transparency and detection robustness when embedding into recycled paper media.

2.3.1 Introduction

Digital watermarking systems have been widely investigated and proposed to address the current increasing trend of document exchanging over the Internet and associated applications. Hardcopy communication is particularly concerned with applications of digital embedding for printed documents. To achieve communication capability, the technique has to deal with distortions introduced by the printing and scanning processes, paper and ink aging [17] and also due to the to user manipulations of printed documents such as rotation [7], spilling, smudging and cropping.

Different approaches to convey information through paper have been proposed which include message embedding into images and/or text of the document [2, 25]. In the literature, Levy and Shaked [13] proposed a hardcopy communication scheme based on DFT domain by hiding information within chosen mid-frequency ranges for transparency. Other work proposed by Varna et al. [24] proposed a technique in the time domain aiming hardcopy communication through text documents. As a result, the approach offers robustness to multiple photocopies. The technique coined Cryptoglyph by AlpVision [1] also proposes a transparent hardcopy approach. When dealing with laser printers, the aforementioned approaches to convey information by employing halftoning modulation can be of interest [19].

This section provides a hardcopy communication technique employing spatial additive embedding. It employs M-ary modulation and achieves a payload over 100 bytes per document page as well as high robustness to the print-scan channel distortions. The robustness is defined as the ability to recover the embedded information after the marked host signal is submitted to the distortions existing in the print and scan channel. The proposed approach was designed to be independent of the document components (graphics, drawings, text, images) as it is able to embed information even in a blank paper (empty page). Many modulation approaches described above are not capable of embedding information into white paper as their associated payload capacity is proportional to the amount and size of the components existing in the document.

The section also presents an analysis for the PS channel which brings insights for understanding and designing a high payload and robust hardcopy communication system. As a result, the approach can be optimized for robustness when dealing with multiple copies by considering the print dot gain, The detection performance can be improved by rejecting interfering components using segmentation based on morphological operations. The provided results illustrate the applicability of the system which can be tuned for practical office document applications including content authentication, author and origin certification, auxiliary channel communication, copy control, etc.

In the following we provide the system analysis and the model for the print and scan channel in Sect. 2.3.2, as important tradeoffs are discussed in Sect. 2.3.3, the component segmentation is discussed in Sect. 2.3.4. Section 2.3.5 provides results and experiments to illustrate the detection performance as Sect. 2.3.6 concludes this section about robustness to multiple photocopies.

2.3.2   Information Encoding and Modulation for the Print and Scan Channel

In this section we describe the models used to investigate and improve performance of the hardcopy communication system presented in Fig. 2.1.

The message or information represented by a string of bits is encoded into a binary B/W (black and white) sequence W by using M-ary, CDM, and TDM modulation processes [15]. This sequence W is then embedded into the digital document X, resulting in the digital marked document:

$$X_W = X + W \tag{2.1}$$

The marked digital document is printed, originating a distorted analogue image:

$$X_{PW} = (X_W \circledast H_P) + \eta_P \tag{2.2}$$

where η_P indicates noise due to the paper texture, dirty or color of the marked document background and H_P indicates the impulse response that models the printer ink spreading. Notice we are assuming a linear model for these distortions where the symbol $\circledast$ indicates linear convolution. This analogue document undergoes external analogue distortions, resulting in a document to be scanned, X_{PWE}.

The document X_{PWE} is then scanned, resulting in the digital document:

$$X_D = X_{PWE} \circledast H_S + \eta_E \tag{2.3}$$

where η_E indicates the digital sensor electronic noise and the quantization noise due to digitalization. H_S (linearly) models the optics of the scanner and the blurring due to sensor mechanical moving. By employing segmentation (at decoder side) in order to extract document interference due components (characters and graphic symbols modeled as $X \circledast H_P \circledast H_S$), a residual segmentation noise is generated, represented by η_C, is included in the model of the digital document presented to detector:

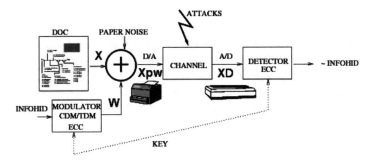

Fig. 2.1 Hardcopy communication scheme

$$X_{DI} = X_D + \eta_C - X \circledast H_P \circledast H_S \tag{2.4}$$

The resulting digital document presented to detector:

$$X_{DI} = (W \circledast H_P + \eta_P) \circledast H_S + \eta_E + \eta_C \tag{2.5}$$

We disregard other external geometrical distortions such as rotation and cropping at this point of the analysis.

Depending on the type of media, different amounts of noise needs to be addressed. For paper media such as standard office or glossy, we may assume that η_P is negligible. However, the paper noise η_P needs to be considered in the model when dealing with recycled paper media which presents a great deal of paper texture and noise. Moreover, in practice [4] the resulting blurring caused by the PS channel can be assumed to be dominated by the optics of the scanner, thus, $H_S \circledast H_P \approx H_S$. The electronic noise is assumed negligible when compared to the residual segmentation noise (in lack of proper segmentation, document interference is a very stronger noise) and the paper noise, thus $\eta_C + \eta_E \approx \eta_C$. Considering these simplifications and that we are employing a segmentation algorithm to extract the components (text, drawings, graphics, etc.), the simplified model becomes $X_{DI} \approx W \circledast H_S + \eta_C$.

We employ normalized zero-lag correlation function defined as:

$$\langle X, Y \rangle = \frac{1}{N} \sum_{i=1,\dots,N} X_i Y_i \tag{2.6}$$

to detect the presence of a sequence W in a document assuming hypothesis H1 (the mark is present), the resulting detection metric becomes:

$$\rho_1 = \langle W, X_{DI} \rangle = \langle W, W \circledast H_S \rangle + \langle W, \eta_C \rangle \tag{2.7}$$

On the other hand, for the hypothesis H0: the W mark is not present. In this case, the resulting detection metric becomes:

$$\rho_0 = \langle W, X_{DI} \rangle = \langle W, \eta_C \rangle \tag{2.8}$$

Notice that the chosen zero lag correlation metric performs a sum of N samples of a signal with some unknown probability distribution function (pdf). Moreover, as the Central Limit Theorem assures that for large N, a sum of disturbances (as in the zero-lag correlation metric) will present a normal distribution. This approximation has been verified for the zero-lag correlation metric since our problem requires a sufficient large N: size of the mark. Experiments also indicate that noise η_C follows a uniform distribution and plays a role by decreasing the detection probability according to our analysis. We may consider that the paper and electronic noises are negligible when dealing with standard or glossy white paper media. However, when dealing with recycled and textured media, the optimal detector is not the zero lag correlation. The optimal detector can be derived by considering the problem of detecting a known signal distorted by a correlated noise. To solve this problem we need to estimate the correlation matrix for the paper noise η_P, as described in [21]. Thus, this model allows us to derive the probability of finding the embedded message in the document by exploiting detection theory either for glossy white or recycled paper media.

By employing the M-ary modulation, the document is marked with a set of NW marks per block region (Code Division Modulation combined with TDM modulation). The M-ary modulation introduces NW mark signals in each block. Whenever the mark W_k is present in the block, the detection metric is found as:

$$\rho_1 = \langle W_k, X_{DI} \rangle = \langle W_k, W \circledast H_S \rangle + \langle W_k, \eta_C \rangle + \sum_{NW-1; i \neq k} \langle W_k, W_i \circledast H_S \rangle \qquad (2.9)$$

However, when the mark W_k is not present in the block:

$$\rho_0 = \langle W_k, X_{DI} \rangle = \langle W, \eta_C \rangle + \sum_{NW; i \neq k} \langle W_k, W_i \circledast H_S \rangle \qquad (2.10)$$

Similar to our previous analyses in [4, 15], the resulting detection probability error can be found as:

$$Pe = PFP + PFN \qquad (2.11)$$

$$Pe = \frac{P_0}{2} erfc \left(\frac{\tau - \mu_{\rho_0}}{\sqrt{2\sigma_{\rho_0}^2}} \right) + \frac{P_1}{2} erfc \left(\frac{\mu_{\rho_1} - \tau}{\sqrt{2\sigma_{\rho_1}^2}} \right) \qquad (2.12)$$

The optimal detection threshold τ can be derived for normal distributions. The parameters ($\mu_{\rho_i}, \sigma_{\rho_i}, i = 0, 1$) can be estimated from the data following the previous assumptions. These parameters can be changed by design as follows to improve detection. The PFN (false negative probability) and PFP (false positive probability) can be reduced by using a small number of NW marks per block region while this reduces the payload. Approaches to estimate H_S (known as matched filter) are used to improve detection performance (ρ_1 depends on $\langle W_k, W \circledast H_S \rangle$) by correlating X_{DI} with $W_k \circledast H_S$ instead of W_k. This procedure takes into account the print dot gain generated by the printer and the scanner optics. When multiple copies are taken from the document, this procedure needs to be iterated as many times as the number of copies, thus enlarging the dot patterns (according to resulting print dot gain) and improving the resulting correlation. Better performance is also achieved by designing marks with maximum orthogonality to minimize cross-correlations:

$$\sum_{NW;i\neq k} \langle W_k, W_i \circledast H_S \rangle \tag{2.13}$$

Similar optimization is derived in [15]. In case of dealing with non-malicious geometrical distortions such as non-intentional cropping and small rotations, the proposed method can be used by employing special visual cues for synchronization and error correction approaches as described in [7].

As indicated, the detection is achieved by correlating the reference marks (and decoding with M-ary decoder), w, with the marked and segmented document X_{DI}. Performing the correlation in the time domain is not computationally feasible due to the large amount of pixels in a scanned document:

$$c[m] = \sum_{n=-\infty}^{\infty} w[n]X_{DI}[n+m] = w[n] \circledast X_{DI}[-n] \tag{2.14}$$

Thus, we employ an exponentially faster correlation computation in the frequency domain exploiting the correlation theorem:

$$c[m] = FFT^{-1} \left(FFT(w_e[n])FFT(X_{DI}[-n]) \right) \tag{2.15}$$

where $w_e[n]$ is the mark extended with zeros to the size of $X_{DI}[n]$ and FFT stands for Fast Fourier Transform.

The M-ary modeling of information is constrained to the number of detections allowed (time complexity), ND, number of regions (NR), number of watermarks per region (NW), total number of marks (NT) and number of embedding bits (payload). By allowing NW marks per region (not necessary the same for all regions) from a database of NT marks, we can achieve a payload of

$$log_2(C_{NW}^{NT}) \times NR \tag{2.16}$$

bits, where

$$C_{NW}^{NT} = \frac{NT!}{(NT - NW)! \times NW!} \tag{2.17}$$

The resulting total number of required individual detections to decode the message is $ND = NT$. For instance, to achieve payload of 100 bytes using 4 marks per region, from a database with a total of 50 marks, we need to mark 44 (800/18) regions. It requires only 50 frequency domain correlations in order to decode the message.

2.3.3   Security and Transparency Versus Robustness Tradeoff

Our experiments indicate that sparse black dots can be properly hidden on recycled substrates as shown in Fig. 2.2b. High payload hardcopy communication with black patterns on white regular paper generate visible artifacts as shown in Fig. 2.2a. The recycled paper has lots of paper noise and it degrades the detection performance as shown in Fig. 2.3. On the other hand, yellow dots provide the most transparent printed patterns, as perceived by human eyes, due to the surrounding white. However, yellow dots do not survive black-and-white photocopies which limits considerably the applications of the proposed hardcopy watermarking system. Therefore, when robustness to multiple copies are required, yellow dots cannot be used as they easily fade away after a B/W copy. Thus, in this case, we need to trade the transparency by robustness to multiple B/W copies using black dots to modulate the document. For the authentication applications we are interested in a visible but not intrusive pattern in the background. These visible patterns should be not a problem since the user will not intentionally destroy it as usually he/she is interested in keeping the document authenticated. We only need to make sure that an attacker is unable to reproduce a valid pattern in order to fake the authentication by replacing the background. This can be done using a secure protocol for keeping safe the keys to enable us to generate valid patterns and keep away attackers.

(a)   **(b)**

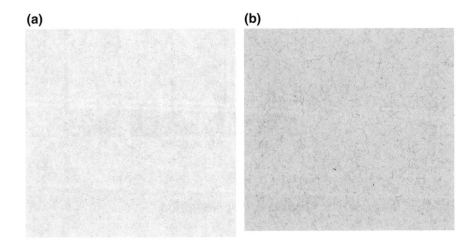

Fig. 2.2 Resulting perceptual impact: (**a**) four patterns embedded into white paper and (**b**) four patterns embedded into recycled paper. Notice that recycled paper provides higher perceptual transparency but more paper noise η_P, and it results in bigger PFN and PFP

Fig. 2.3 Resulting detection performance: (**a**) detection by correlation for four patterns embedded into white paper and (**b**) detection by correlation for four patterns embedded into recycled paper. Notice that recycled paper provides higher perceptual transparency with more paper noise η_P, and it results in bigger PFN and PFP

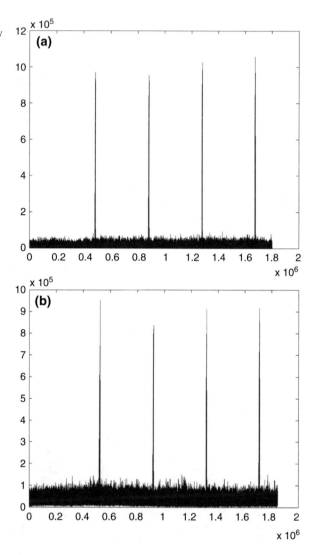

2.3.4 Component Rejection by Segmentation

As the embedded mark does not depend on the document components (text, image, drawing and graphics), most of the existing components can be segmented out before detection. When embedding with additive sparse black dots, we can reject components by size (connected components) or by using morphological filtering. The use of high pass filtering to remove the large components or other ineffective segmentation leaves a great amount of segmentation noise, the noise η_C described in Sect. 2.3.2. The performance of the proposed spatial morphological segmentation approach is shown in Fig. 2.4b. The segmentation removes components larger than the dot patterns prior the detection by correlation.

2.3.5   Experiments

Figure 2.4 illustrates the splitting of the document into 80 regions. Each region will be embedded with four patterns using the M-ary modulation. In these experiments we are using standard office laser printer and scanning at 300 dpi.

Figure 2.5 illustrates the original watermarked document and the scanned marked document modulated by black dots after one pass over the PS channel. The employed strong modulation assures proper message detection robust to four copies by trading-off on the transparency property of the mark as illustrated in Fig. 2.6. This is a useful feature of the proposed algorithm and the performance can be tuned by the variables NW and NT.

In Fig. 2.7 we present experiments to evaluate the resulting PFN and PFP for a given a detection threshold τ, using black dot modulation and multiple copies, the resulting false positive probability is $\frac{P_0}{2}erfc\left(\frac{\tau-\mu_{\rho_0}}{\sqrt{2\sigma_{\rho_0}^2}}\right)$. The resulting false negative probability is $\frac{P_1}{2}erfc\left(\frac{\mu_{\rho_1}-\tau}{\sqrt{2\sigma_{\rho_1}^2}}\right)$.

2.3.6   Conclusions

This section describes a hardcopy communication system based on M-ary, TDM and CDM modulations. We modeled the print and scan channel and based on this study a framework is given for optimizing the system for robustness to multiple copies. The modulation is independent of the document components and thus the use of a proper component segmentation improves the performance significantly. The results clearly illustrate the performance of the proposed system. The provided hardcopy communication system can be combined with other modulation system based on character and word modulations [2].

The next section is dedicated to exploit the properties of color inkjet print channels and also to propose an approach to efficiently convey information over this challenging channel.

2.4   Hardcopy Communication for Inkjet Channels

This section describes novel approaches to achieve robust communication over inkjet print-and-scan (IPS) color channels. The IPS color channel poses even greater challenges than the laser print-and-scan channel due to the resulting mixing and spreading of the ink dots. We discuss a novel informed coding and two host color rejection approaches. One is based on a novel color rejection and the other on a whitening filter in order to deal with the aforementioned inkjet printer distortions. A substitutive spatial domain embedding is employed to enable robustness

(a)

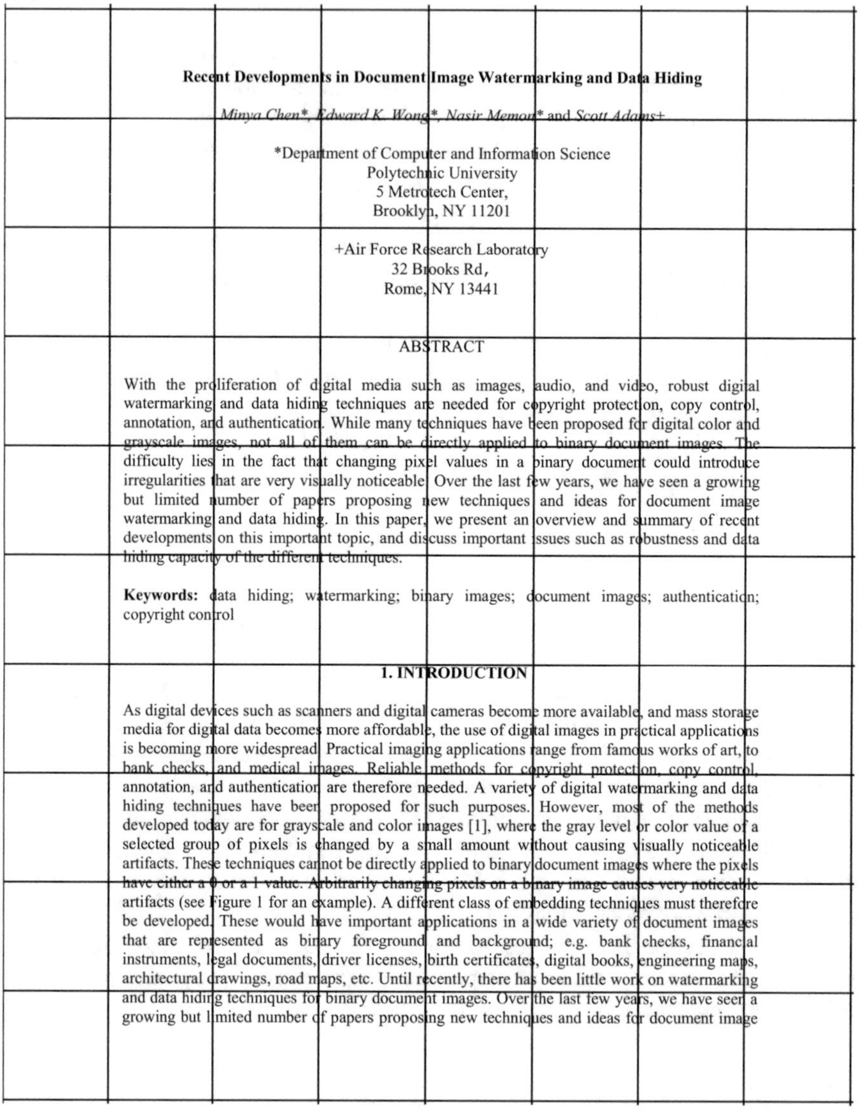

Recent Developments in Document Image Watermarking and Data Hiding

Minya Chen*, Edward K. Wong*, Nasir Memon* and Scott Adams+

*Department of Computer and Information Science
Polytechnic University
5 Metrotech Center,
Brooklyn, NY 11201

+Air Force Research Laboratory
32 Brooks Rd,
Rome, NY 13441

ABSTRACT

With the proliferation of digital media such as images, audio, and video, robust digital watermarking and data hiding techniques are needed for copyright protection, copy control, annotation, and authentication. While many techniques have been proposed for digital color and grayscale images, not all of them can be directly applied to binary document images. The difficulty lies in the fact that changing pixel values in a binary document could introduce irregularities that are very visually noticeable. Over the last few years, we have seen a growing but limited number of papers proposing new techniques and ideas for document image watermarking and data hiding. In this paper, we present an overview and summary of recent developments on this important topic, and discuss important issues such as robustness and data hiding capacity of the different techniques.

Keywords: data hiding; watermarking; binary images; document images; authentication; copyright control

1. INTRODUCTION

As digital devices such as scanners and digital cameras become more available, and mass storage media for digital data becomes more affordable, the use of digital images in practical applications is becoming more widespread. Practical imaging applications range from famous works of art, to bank checks, and medical images. Reliable methods for copyright protection, copy control, annotation, and authentication are therefore needed. A variety of digital watermarking and data hiding techniques have been proposed for such purposes. However, most of the methods developed today are for grayscale and color images [1], where the gray level or color value of a selected group of pixels is changed by a small amount without causing visually noticeable artifacts. These techniques cannot be directly applied to binary document images where the pixels have either a 0 or a 1 value. Arbitrarily changing pixels on a binary image causes very noticeable artifacts (see Figure 1 for an example). A different class of embedding techniques must therefore be developed. These would have important applications in a wide variety of document images that are represented as binary foreground and background; e.g. bank checks, financial instruments, legal documents, driver licenses, birth certificates, digital books, engineering maps, architectural drawings, road maps, etc. Until recently, there has been little work on watermarking and data hiding techniques for binary document images. Over the last few years, we have seen a growing but limited number of papers proposing new techniques and ideas for document image

Fig. 2.4 **(a)** Digital document is divided into 80 regions to be marked with TDM. **(b)** Segmented printed watermarked document by rejecting large components. Embedded dots are hardly seen and the proposed segmentation has a small η_C. One needs to zoom in to be able to notice the noise and the embedded dots

(b)

Fig. 2.4 (continued)

(a)

With the proliferation of digital media such as images, audio, and video, robust digital watermarking and data hiding techniques are needed for copyright protection, copy control, annotation, and authentication. While many techniques have been proposed for digital color and grayscale images, not all of them can be directly applied to binary document images. The difficulty lies in the fact that changing pixel values in a binary document could introduce irregularities that are very visually noticeable. Over the last few years, we have seen a growing but limited number of papers proposing new techniques and ideas for document image watermarking and data hiding. In this paper, we present an overview and summary of recent developments on this important topic, and discuss important issues such as robustness and data hiding capacity of the different techniques.

(b)

With the proliferation of digital media such as images, audio, and video, robust digital watermarking and data hiding techniques are needed for copyright protection, copy control, annotation, and authentication. While many techniques have been proposed for digital color and grayscale images, not all of them can be directly applied to binary document images. The difficulty lies in the fact that changing pixel values in a binary document could introduce irregularities that are very visually noticeable. Over the last few years, we have seen a growing but limited number of papers proposing new techniques and ideas for document image watermarking and data hiding. In this paper, we present an overview and summary of recent developments on this important topic, and discuss important issues such as robustness and data hiding capacity of the different techniques.

Fig. 2.5 (**a**) Digital marked document. (**b**) Scanned marked document after one copy

optimization using the proposed informed coding. Analyses and examples are provided to evaluate the performance. It is shown the enhancement on robustness and transparency achieved by the proposed approaches.

Inkjet printers are usually much affordable than color laser printers, as a result, they are widely available in many countries in development. Many security (as in tracking the origin/author of a document) and authentication (as in authenticating contents of a document for product piracy or fraud determent) applications deploy hardcopy communication techniques that are properly designed for this specific type of printer technology, namely, inkjet printers. For inkjet print-and-scan channels, however, additional techniques to the aforementioned approaches are required in order to deal with the stronger channel distortions due to the IPS ink mixing and spreading. Some work in this direction is proposed in [16] and the informed coding approach is inspired by the work of Professor Max H. Costa [6] named "Writing on Dirty Paper". The informed coding approach has been exploited by [18] for single channel (monochromatic) image modulation achieving very high payload for noise, filtered and compressed channels. However, that approach is not designed to convey information on color host images over the IPS color channel.

This section provides detailed discussions and analyses on the robustness and detection improvements due to the novel proposed informed coding and color rejection techniques designed to deal with color IPS channels.

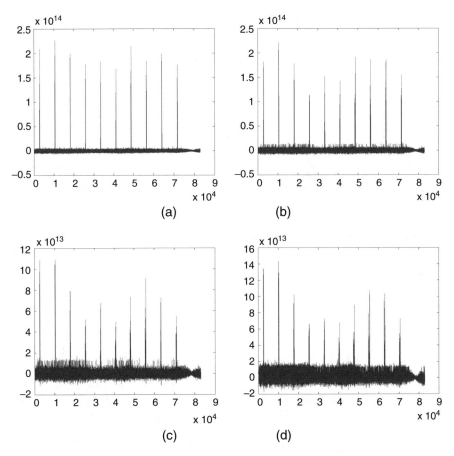

Fig. 2.6 Resulting correlation performance after (**a**) one copy, (**b**) two copies, (**c**) three copies and (**d**) four copies

2.4.1 Information Embedding

Let us encode the desired message into an *m*-bit string. This message needs to be embedded and conveyed by a digital color image I, which later on is printed and dispatched as an inkjet printed media. Our embedding approach employs a set of K color dot patterns (a randomly generated sparse matrix of color dots) from a set of N digital patterns, $P_i, i = 1, \ldots, N, (N \geq K)$. This set of K patterns is uniquely represented by an unordered set $S = \{k_1, k_2, \ldots, k_K\}$, where $k_i \neq k_j$ for $i \neq j$. The resulting embedding provides the following marked document I_w:

$$I_w = I \otimes \sum_{i \in S} P_i \qquad (2.18)$$

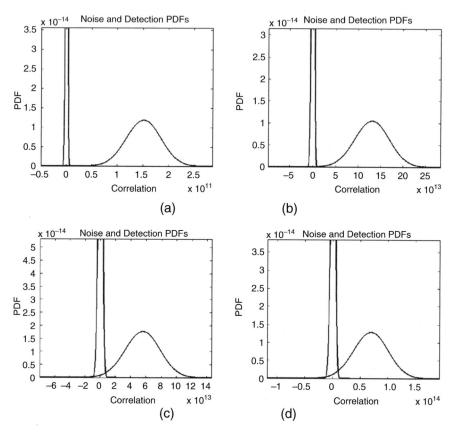

Fig. 2.7 Resulting detection statistics performance after (**a**) one copy, (**b**) two copies, (**c**) three copies and (**d**) four copies

where the operation $\otimes$ represents an image pixel substitution wherever a color dot pattern exists. This embedding differs considerably from the traditional additive embedding. In this case, the pixels of the image I are replaced by the pixels of the color pattern, whenever color dots exist, as illustrated in the left side of Fig. 2.8.

This approach results in a very transparent embedding when employing sparse patterns with small size square dots. The resulting perceptual impact is low because the inkjet printing process, due to ink spreading, mix around the small size embedding dots and hides the embedding pattern as illustrated in the right side of Fig. 2.8. The transparency is achieved in this case by selecting dot sizes smaller than 6×6 pixels in resolutions of 600 dpi/ppi for printing and scanning. In the experiments, in Sect. 2.4.6, we provide performance evaluations employing small dot sizes of 4×4 pixels for each pattern at 600 dpi/ppi resolutions for printing and scanning. For many applications, such as authentication, the transparency requirement may be relaxed. In contrast, the resulting impact of employing 2D barcodes, may be too high for many applications. The proposed substitutive approach is recommended in this

Fig. 2.8 The digital marked image and the image after the color IPS channel. The digital marked image detail (with 1150×850 pixels, corresponding in a real size of 1.9 in $\times$ 1.4 in) with a cyan pattern. The original image, printed and scanned at 600 dpi/ppi, has 3200×2400 pixels. The digital domain is used only for embedding as the distribution media is the printed version where the resulting embedding transparency is high. On the right is shown the printed and scanned marked image with a cyan pattern

case because the use of additive embedding instead, with the same encoding, would require a greater amount of energy and therefore would result in lesser transparency considering the inkjet ink properties.

2.4.2   On the Choice of Patterns for Informed Coding

Our experiments indicate that the ink spreading of the dots differs depending on the color of the embedding pattern dots, k, and also on the colors of the pixels surrounding the embedded dots in the host image, as shown in Fig. 2.9. This occurs due to ink chemical properties of the inkjet printer cartridges and also depends on the paper media. Therefore, the resulting detection performance based on correlation is considerably higher for certain color combinations of the embedding patterns and the host backgrounds. Based on these findings, we propose to employ the correlation operation as the robustness metric. It is used for selecting the best pattern, consisting of dots of one unique color, from L alternative patterns. The chosen one is selected by considering the pixel colors of the image host background under the pattern dots.

Fig. 2.9 Dot spreading on
different backgrounds due to
the inkjet ink properties.
Detail (zoom) of the color dot
pattern ink spreading and
mixing for different color
backgrounds

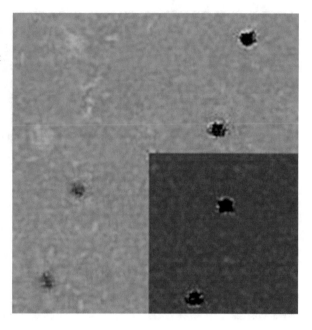

In order to illustrate this choice, the robustness estimates are illustrated in
Fig. 2.10 for four background colors using an HP4280 multifunction inkjet printer.
These estimates indicate that higher robustness is achieved by embedding magenta
pattern dots over a cyan image background rather than over a yellow background. To
estimate this robustness, $R(k, b)$, of embedding a pattern of color k in a background
of color b, a training step is required for a given printing system (printer, inkjet
cartridges and scanner). The training estimates all available combinations of colors
of the patterns k and backgrounds b. One approach to speed up this step is to employ
fewer background representative colors, by clustering sets of background colors
using Euclidean distance. It also helps to reduce the complexity of the informed
coding step. Thus, for a given color k, a pattern P_{i*k} from a set of L equivalent
patterns is chosen at embedding by maximizing

$$P_{i*k} = \max_{i=1,L} \Phi\{ P_{ik}, I_w\} \tag{2.19}$$

where $\Phi\{ P_{ik}, I_w\}$ represents the average robustness for all D pattern dots, after IPS
channel, between the D color dots of pattern P_{ik} and the U pixels surrounding these
dots in the host image. Thus,

$$\Phi\{ P_{ik}, I_w\} = \frac{1}{UD} \sum_{d=1}^{D} \sum_{[r,s]\in\mathcal{N}_d} R(k, Color(I_w[r, s])) \tag{2.20}$$

where $[r, s] \in \mathcal{N}_d$ represents the set of U pixels at the neighborhood of the embedded
dot d and $Color(I_w[r, s])$ is the color of the marked image at location $[r, s]$.

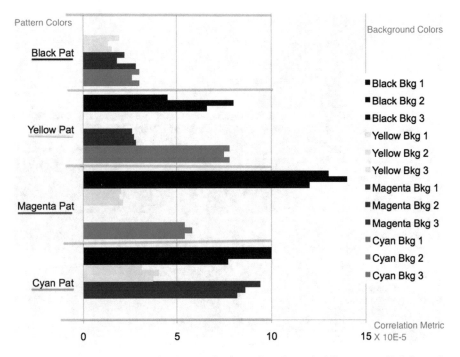

Fig. 2.10 Expected robustness for three realizations of a color embedding pattern (Pat) for each (CMYK) color background (Bkg)

In order to implement the informed coding approach, L alternative patterns for each of the N patterns are created. The total number of patterns is $N{\times}L$. The message is informed encoded by searching for the best set of K embedding patterns by employing our robustness metric. The total number of required individual detections to decode the message is $ND = N \times L$. As we propose to use L alternative patterns, for each of the N patterns, which convey the same message, both encoder and decoder must share a secret key ϕ in order to generate the same set of LN patterns.

The informed coding approach allows an information payload (m bits) determined by the number K of color patterns per region, the total number N of patterns and the number N_R of embedding regions (time division modulation). By employing K patterns per region from a set of N patterns and considering N_R embedding regions, the resulting payload of m bits is:

$$m \leq N_R \, \log_2 \binom{N}{K} \tag{2.21}$$

where $\binom{N}{K} = N!/[K!(N-K)!]$

Fig. 2.11 Host images without marks after IPS color channel. Original (no marks) images of size $1.5 \, \text{in}^2$ printed at 600 dpi and scanned at 600 ppi in HP5580

2.4.3 Color Rejection Approach to Reduce Host Interference

As each pattern in a region is defined with a unique color k, the detection metrics in (2.23) is computed for this pattern considering only this k color channel. The rejection of pixels is used to remove the other colors in the received image by considering a statistical distance defined as follows. Let us represent the pixel of color k as a vector with CMYK color components: $\boldsymbol{B} = [B_c \, B_m \, B_y \, B_k]^T$. Assume this pixel color is a random variable distributed as $\boldsymbol{B} \sim N(\boldsymbol{\mu}_k, \boldsymbol{C}_k)$. After transmitting Z pixels of such color k over a given IPS channel, we estimate the mean vector and covariance matrix as $\boldsymbol{\mu}_k = \frac{1}{Z} \sum_i \boldsymbol{B}_i$ and $\boldsymbol{C}_k = \frac{1}{Z} \sum_i \boldsymbol{B}_i \boldsymbol{B}_i^T - \boldsymbol{\mu}_k \boldsymbol{\mu}_k^T$. Therefore, when testing for a pattern of color k, an unknown color pixel $\boldsymbol{X}$ of the received image is accepted only if the Mahalanobis distance to the color k, defined by

$$d_{M_k}(\boldsymbol{X}) = \sqrt{(\boldsymbol{X} - \boldsymbol{\mu}_k)^T \boldsymbol{C}_k^{-1}(\boldsymbol{X} - \boldsymbol{\mu}_k)}, \qquad (2.22)$$

is smaller than to the distance to other colors: $d_{M_k}(\boldsymbol{X}) < d_{M_i}(\boldsymbol{X}), i \neq k$. The proposed criterion of rejection is optimal for Normal distributed pixels and the validity of this assumption is verified in the experiments. As an example, let us embed $K = 4$ patterns in a region, each with one unique color from CMYK. Thus, in order to

Fig. 2.12 Marked images after IPS color channel. Printed and marked images indicated high transparency when viewed from a distance of 10 in. There are less than 1% of pixels marked as cyan dots of size of 4 × 4 pixels, which are barely seen by naked eye. Only with digital zoom is possible to notice the patterns

detect the $k = Y$ (yellow) pattern, the algorithm rejects the other (CMK) colors generating the modified image I_{wR} before correlation.

2.4.4   Pattern Detection Followed by Message Decoding

After the embedding with K color patterns by region, the marked color image is printed using an inkjet printer and the marked document is distributed. This document is then digitized using a scanner before the message decoding step. The detection of a pattern P_{ik} is performed using the correlation operation performed in the frequency domain (for speed), after rejecting the other colors $\neq k$, as described above. The correlation is computed between the observed marked image after the print-scan channel and the N_D patterns in the set, which can be generated at decoder using a shared secret key ϕ.

These images (received digitized document with the other colors rejected and the color pattern) are converted to the HVS (Hue, Value and Saturation) color model. Before applying the correlation, a $LoG[m, n]$ whitening filter (Laplacian of Gaussian

[8]) of dimension 3×3 is applied in order to decorrelate the host signal (document with colors $\neq k$ reject). The entire operation consists of an average correlation over the channels of the document I_{wR} and known pattern P_i represented in HVS color model:

$$C_i = \frac{1}{3} \sum_{z=H,V,S} \mathscr{F}^{-1}\{\mathscr{F}(I_{wR_z}[m,n] \circledast LoG[m,n]) \cdot \qquad (2.23)$$

$$\mathscr{F}(P_{ikz}[-m,-n] \circledast LoG[-m,-n])\}$$

where $\mathscr{F}(\cdot)$ and $\mathscr{F}^{-1}(\cdot)$ are the fast 2D direct and inverse discrete Fourier transforms, respectively. The 2D linear convolution, indicated by the operator $\circledast$, performs the whitening filtering over the image and the pattern. The K indexes of patterns P_i are the set S of indexes required to decode the message m. They correspond to the K highest correlations C_i, and are found after LN detections.

Consider a case where $K = 4$ CMYK colors and redundancy factor of $L = 3$ are chosen. The process requires LN correlations from (2.23) for each of the $K = 4$ colors. The indexes of the K patterns with highest correlation peak value for each color, the set S, are associated by a look-up table to the message m. The decoder will follow the same encoder assignment of message and indexes after it constructs the LN patterns using the key ϕ.

2.4.5 Analysis of the Proposed Detection Metric

The Central Limit theorem states that a large sum of independent small disturbances tends to follow a Normal distribution. Thus, by modeling the detection metric (correlation Ci in (2.23)) as $\sim N(\mu, \sigma^2)$, we find the probability of missing a pattern $P_P(\tau)$ by

$$P_P(\tau) = P_{FN}(\tau) + P_{FP}(\tau) = \qquad (2.24)$$

$$= \frac{P_1}{\sqrt{\pi}} \int_{\frac{\mu_1-\tau}{\sqrt{2\sigma_1^2}}}^{\infty} exp(-t^2)dt + \frac{P_0}{\sqrt{\pi}} \int_{\frac{\tau-\mu_0}{\sqrt{2\sigma_0^2}}}^{\infty} exp(-t^2)dt$$

where $P_{0,1}, \mu_{0,1}, \sigma_{0,1}^2$ parameters are estimated from data and are, respectively, the prior probabilities, means and variances of the detection statistics of regions with no pattern (unmarked hypothesis H_0) and regions with pattern (marked hypothesis H_1). The optimal detection threshold τ is determined and applied to decide the hypotheses based on the observed metric $C_i \overset{H_0}{\underset{H_1}{\lessgtr}} \tau$ by solving

$$(\sigma_0^2 - \sigma_1^2)\tau^2 + 2(\mu_0\sigma_1^2 - \mu_1\sigma_0^2)\tau + \qquad (2.25)$$

$$+\sigma_0^2\mu_1^2 - \sigma_1^2\mu_0^2 + 2\sigma_0^2\sigma_1^2 \ln(\frac{\sigma_1 P_0}{\sigma_0 P_1}) = 0$$

which minimizes the probability of missing a pattern, $P_P(\tau)$. Henceforth, as LN patterns are tested in order to find the K embedded patterns in each one of the N_R regions, the estimated probability of missing the entire message, Pe, is

$$Pe(\tau) = N_R((1 - (1 - P_{FN}(\tau))^K) + (1 - (1 - P_{FP}(\tau))^{LN})) \qquad (2.26)$$

2.4.6 Experiments

The improvements on detection performance are illustrated in Figs. 2.13 and 2.14 providing a huge increase of 70% in μ_1 due to the informed coding after the IPS channel for a chosen pattern P_{i*k} in a given background. Clearly, the detection performance is superior when proper embedding patterns are defined at embedding using the informed coding approach. Moreover, by employing the color rejection, the correlation statistics μ_1 is increased by 15% while de deviation σ_1 is decreased by 50%, providing an huge performance improvement on detection.

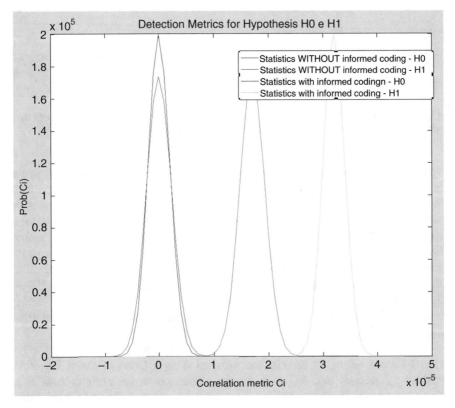

Fig. 2.13 Performance of the informed coding approach. The correlation performance **with and without** informed coding: the mean μ_1 for the hypothesis H_1 is improved by 70%

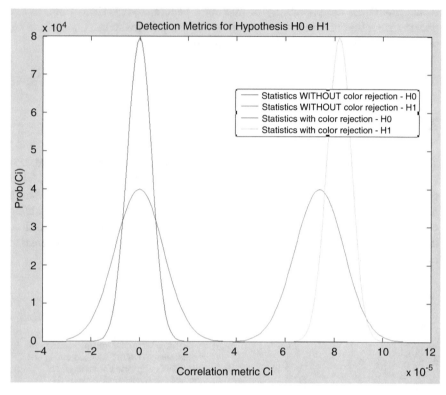

Fig. 2.14 Performance of the color rejection approach. The correlation performance **with and without** color rejection: the μ_1 is improved by 15% and $\sigma_{0,1}$ are decreased by 50%

Considering a payload of 1035 bits, at least 23 bits per region need to be embedded into a region using $NR = 45$ regions, a color host image larger than 2100×2100 pixels and $K = 4$ patterns per region. According to (2.21), it would be necessary to detect $N = 140$ patterns with $L = 3$ alternatives each (informed coding). By estimating the distribution parameters from IPS experiments, the resulting probability $P_{FN}(\tau)$ is about 4×10^{-10} and $P_{FP}(\tau)$ is about 3.9×10^{-10} for a given image. For this payload the estimated probability of missing the entire message is $Pe(\tau) = 45((1 - (1 - 4 \times 10^{-10})^4) + (1 - (1 - 3.9 \times 10^{-10})^{140 \times 3})) = 7.4 \times 10^{-6}$. This estimation shows that the approaches provide a very robust embedding for the IPS channel, which can be further improved by using an error correction code.

The performance is validated by computing the correlation metrics statistics (μ, σ) from a set of 50 marked and scanned images (sizes of about $1.5\,\text{in}^2$ at 600 ppi/dpi resolutions), following by the estimation of the error probabilities. Some images are illustrated in Figs. 2.11 and 2.12. After employing the proposed color rejection and informed coding approach, the lowest performance case for

$[Pe(\tau), P_{FN}(\tau), P_{FP}(\tau)]$ is improved from $[2.4 \times 10^{-5}, 2.1 \times 10^{-7}, 5.5 \times 10^{-8}]$, respectively, to $[4.3 \times 10^{-6}, 4.4 \times 10^{-8}, 9.7 \times 10^{-9}]$. The results indicated a consistent improvement, for different messages and multifunctional printers/scanners (HP5580 and HP4280), of at least five times in probability of detection even for the worst of the 50 cases.

Prior to printing the image, the resulting Peak Signal-to-Watermark Ratio, $PSWR = 20log_{10}\left(\frac{255 \times W \times H}{\sqrt{|I - I_w|^2}}\right)$, is very high, $PSWR = 45\,\mathrm{dB}$, on average, where W and H are respectively the width and height of the images. The Structural Similarity (SSIM) index [26] is also high prior printing, $SSIM = 0.96$ in average.

Perceptual evaluation from 15 users indicates a transparent embedding to naked eyes from a normal distance (10 in) to the printed page using about 400 dots of 4×4 pixels size per pattern of 300×300 pixels at 600 ppi/dpi resolutions. IPS channel hides those dots quite well due to ink spreading and mixing, as illustrated in Figs. 2.8, 2.9, 2.11 and 2.12. This setup provides a payload of 100 bits/in^2 at 600 ppi/dpi for IPS color channels with a small probability of missing the message ($\sim 10^{-5}$) and good transparency, a performance very competitive to hardcopy communication techniques discussed in Sect. 2.4.

In all the experiments, a careful placement of the printed document in the scanbed resulted in a well aligned image. Notice that the correlation method is robust to any degree of translation; however, the performance may be affected if rotation occurs. In this case we apply the following automatic method based on a coarse alignment followed by a search rotation method to achieve a finer alignment: The corners and boundaries of the image are detected and a rotation is performed on the image to achieve a coarse alignment. Next a fine search is performed aimed to improve the alignment. This is achieved by computing the correlation of some blocks and use this information to select the angle that provided the highest correlation. This approach does not require any additional visual cues or blocks specially designed to recover synchronism. The approach has the drawback of requiring extra computational time to find the best rotation angle from a range of few degrees after the coarse alignment. Figure 2.15 illustrates the correlation performance dependent on the rotation angle.

Transparency, payload, decoding speed and robustness are adjustable by using a different set of parameters N, K, N_R, L, dot width and number of dots per pattern for a given number m of embedding bits. Higher payload is achievable by increasing N with some impact on computational complexity and robustness. According to (2.20) the embedding has complexity proportional to the product $N_R KLUD$. In the experiments, the embedding require a mean of 3.5 min for images of size 800×800 pixels while the decoding using (2.23) required a mean of 40 s using a non-optimized program.

The transparency and detection performance can be adjusted by changing the number of dots used in the patterns. The detection is improved by increasing the number of dots while the transparency is decreased with more dots per pattern. The tradeoff between transparency and detection depends on a particular application.

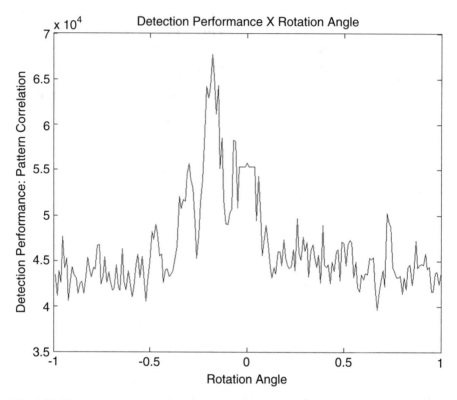

Fig. 2.15 The correlation performance for a range of rotation angles. This approach enables to determine the best rotation angle automatically making the approach robust to angle rotation at the scanning process. The detection method based on correlation is already naturally robust to translation

Notice that our approach filter out most of the image contents before testing for a pattern by rejecting image pixels of colors more distant than the pattern color. Thus, both detection and transparency depends mostly on the number of dots pattern used to embed. The embedding by distributing the dots uniformly over the image, as illustrated in Fig. 2.16, results in a correlation peak of 19×10^{-4} for the cyan pattern. By reducing the number of dots we achieve a higher transparency and lower detection, however, in this case, we are still embedding dots in smooth regions which are much more perceived than the dots allocated in active regions (regions with texture, noise, edges, etc.).

To address this issue of perceptual impact on smooth region we propose to estimate the active regions by using a combination of linear filters (a Laplacian of a Gaussian filter similar as the one used in (2.23)) and the opening morphological operation to create a mask of smooth regions that should be preserved in the embedding. We define a low pass filter $lp[m, n]$ using a mask of 5×5 coefficients

Fig. 2.16 Uniformly distributing dot results in less transparent embedding with a correlation peak of 19×10^{-4}. Zoom at printed and scanned marked image (the actual size is about half inch wide, see Fig. 2.21) using a multifunctional HP5580 with printer/scanning density of 600 dpi/ppi. Each embedding cyan dot has width of a 4/600 of an inch, practically invisible to naked eyes looking in a printed paper

with values equal to 1/25. A high pass filter $hp[m, n]$ is defined with a mask of 5×5 coefficients with values of -1 except the central coefficient, which is set to 24. The filtered image $I_F[m, n]$ is obtained by applying the filters in sequence to the host image $I[m, n]$:

$$I_F[m, n] = hp[m, n] \circledast \{lp[n] \circledast I[m, n]\} \qquad (2.27)$$

where the operator $\circledast$ represents the discrete convolution. A binary image is created by thresholding the filtered image with a defined value κ and by applying to this image the opening morphological operator [8], $OP(\cdot)$, to generate the final mask, $I_{MASK}[m, n]$, which indicates the smooth regions in the image:

$$I_{MASK}[m, n] = OP(I_F[m, n] < \kappa) \qquad (2.28)$$

where $OP(\cdot)$ uses a square structure element of size 5×5. We propose to embed the dots mostly in the active regions and the number of dots and size of the smooth regions is controlled by the parameter κ. By selecting $\kappa = 0.4$ we achieve a decrease in the number of embedded dots to about 50% as illustrated by the resulting mask in Fig. 2.17. Thus, the remaining dots are allocated in the most active regions as

Fig. 2.17 The $I_{MASK}[m, n]$ defined in (2.28) with $\kappa = 0.4$ to estimate the smooth and active regions. This approach allows the allocations of the pattern dots into the most active regions producing a more perceptual transparent embedding

illustrated in Fig. 2.18. The performance of detection by correlation is reduced from 19×10^{-4} to 8×10^{-4} while the perceptual transparency is highly improved. Thus, the detection performance is proportional to the number of embedding dots embedded, while the transparency depends on the number of dots and also what is the region activity where these dots are embedded.

We validated the proposed techniques with a set of 50 images. We adjusted the κ parameter iteratively to achieve a reduction of 50% of embedding dots which resulted in PSNR of about of 51 dB and a structural similarity index measure SSIM [26] of about 0.97 for all images which is considered a very transparent embedding. Other advanced objective quality metrics are found in [11]. One zoom of an image (after the print-scan channel) of this set is shown in Figs. 2.19 and 2.20 with the resulting correlation performance for the cyan pattern. We tested several images and it turns out that what matters most for perceptual quality and robustness is the amount of pattern dots embedded and the amount of smooth regions in the host image as the proposed technique rejects most of the host image contents, before correlation, except the small regions around the pattern dots. Thus, both transparency and robustness depend mostly on the amount of dots embedded and

Fig. 2.18 Distributing the dots according to the estimated smooth regions and reducing the total number of the dots by 50% based on the $I_{MASK}[m, n]$ defined in (2.28) with $\kappa = 0.4$ for this host image. Zoom at image printed and scanned by the multifunctional printer/scanned HP5580 using 600 dpi/ppi. Each embedding cyan dot has width of a 4/600 of an inch, practically invisible to naked eyes in a printed paper, as in Fig. 2.21

how much smooth is the image and very little on the image host contents. Therefore, the size of the set of images used can be considered big enough to validate the technique performance.

We observe that for active regions, the dots are very well hidden, thus we propose to estimate the active regions and embed most of the dots in those regions aiming transparency. The designed embedding dots present a very small size, about 4/600 of an inch wide, and are practically invisible by looking with naked eyes at paper when printed using inkjet printers, their characteristics also help to spread and mix the colors, increasing the perceptual quality. The perceptual transparency of the resulting marked images were evaluate by asking subjects to look at printed versions instead of looking into a zoom in the digital domain after the scanning using 600 ppi. The resulting perceptual transparency was considered very high by subjects when the embedding employed pattern dots of size 4×4 into color images of size of about 2×2 in or bigger. In Fig. 2.21 the marked image is used to authenticate some contents of the surrounding text. Clearly, the authentication marks are efficiently hidden in the printed document. As the original image is not known, nor the secret key to generate the patterns for a given message, no one would be able to decode the message or suspect that some authentication nor side information is carried over with that small image. Many lasers printers embed yellow dots in

Fig. 2.19 Zoom of the embedded image, each embedding cyan dot has width of a 4/600 of an inch, practically invisible to naked eyes in a printed paper

not-printable regions of the paper in order to convey the printer serial number and/or the date of the printing. Our approach also embeds hidden dots but for inkjet printers and inside of color printed images, with wider applications for both security and authentication.

2.5   Conclusions

The last section provided a discussion on conveying information through printed media, provides an overview on literature and discussed improvements on communication over IPS color channels by using a specific approach that includes informed coding, optimal detection and host rejection techniques. These techniques mitigate the host interference as confirmed by the results and the analyses provided. The detection performance is evaluated with the proposed optimal detection threshold and the results illustrate the significant improvement on probability of detecting a

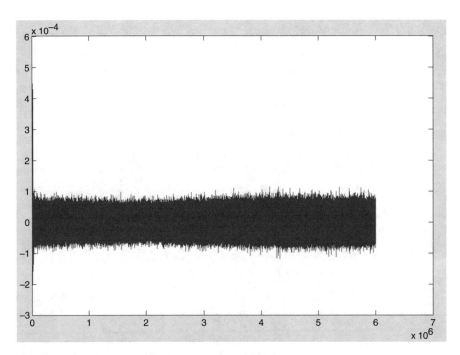

Fig. 2.20 Correlation performance for the cyan pattern, peak of 4.5×10^{-4}. *PSNR* $= 51\,$dB and *SSIM* $= 0.97$

transmitted message over the IPS channel. These approaches improve communication reliability over IPS channels allowing customization for various robustness, transparencies and decoding speed tradeoffs by choosing proper embedding pattern parameters.

This chapter provides a review on hardcopy watermark techniques, discusses two approaches to address communication over PS channels. The first approach provides an encoding for robust authentication of documents after multiple passes. The second approach addresses the challenge problem of conveying information over IPS (Inkjet Print Scan) channel by proposing informed coding, optimal detection and host rejection techniques. These examples of approaches provide guidance for researchers to design new solutions considering the issues addressed and techniques discussed in this chapter.

Next chapter will discuss fundamentals and approaches to convey hidden information by text modulation techniques on printed documents.

Fig. 2.21 The marked image can be used to authenticate some contents of a document. The marks are efficiently hidden from an observer looking at normal view distance

References

1. AlpVision SA, Cryptoglyph digital security solution. (http://www.alpvision.com)
2. P.V.K. Borges, J. Mayer, Text luminance modulation for hardcopy watermarking. Signal Process. **87**, 1754–1771 (2007)
3. P.V.K. Borges, J. Mayer, E. Izquierdo, Robust and transparent color modulation for text data hiding. IEEE Trans. Multimedia **10**, 8 (2008)
4. P.V.K. Borges, J. Mayer, E. Izquierdo, Document image processing for paper side communications. IEEE Trans. Multimedia **10**, 1277–1287 (2008)
5. P. Bulan, G. Sharma, V. Monga, Orientation modulation for data hiding in clustered-dot Halftone prints. IEEE Trans. Image Process. **19**, 8 (2010)
6. M.H.M. Costa, Writing on dirty paper. IEEE Trans. Inf. Theory **IT-29**, 439–441 (1983)
7. F. Deguillaume, S. Voloshynovskiy, T. Pun, Method for the estimation and recovering from general affine transforms, in *Proceedings of SPIE, Electronic Imaging, Security and Watermarking of Multimedia Contents IV*, vol. 4675 (2002), pp. 313–322
8. R.C. Gonzalez, R.E. Woods, *Digital Image Processing* (Prentice Hall, Upper Saddle River, 2001)

9. C. Guo, G. Xu, X. Niu, Y. Yang, Y. Li, A color image watermarking algorithm resistant to print-scan, in *IEEE International Conference on Wireless Communications, Networking and Information Security* (2010)

10. D. He, Q. Sun, A practical print-scan resilient watermarking scheme, in *IEEE International Conference on Image Processing* (2005)

11. H. Hofbauer, A. Uhl, An effective and efficient visual quality index based on local edge gradients, in *3rd European Workshop on Visual Information Processing (EUVIP)* (2011), pp. 162–167

12. A. Keskinarkaus, A. Pramila, T. Seppänen, Image watermarking with a directed periodic pattern to embed multibit messages resilient to print-scan and compound attacks. J. Syst. Softw. **83**, 1715–1725 (2010)

13. A. Levy, D. Shaked, *A Transform Domain Hardcopy Watermarking Scheme* (HP Labs, Haifa, 2001)

14. Q. Li, I.J. Cox, Using perceptual models to improve fidelity and provide resistance to valumetric scaling for quantization index modulation watermarking, in *IEEE Transactions on Information Forensics and Security* (2007), pp. 127–139

15. J. Mayer, J.C.M. Bermudez, Improving robustness *IEEE International Conference on Acoustics, Speech and Signal Processing, ICASSP* (2007)

16. J. Mayer, S. Simske, Informed coding for color hardcopy watermarking, in *8th International Symposium on Image and Signal Processing and Analysis - ISPA* (2013)

17. J. Mayer, J.C.M. Bermudez, A.P. Legg, B.F. Uchoa-Filho, D. Mukherjee, A. Said, R. Samadani, S. Simske, Design of high capacity 3D print codes aiming for robustness to the PS channel and external distortions, in *IEEE Conference on Image Processing* (2009)

18. M.L. Miller, G.J. Doërr, I.J. Cox, Applying informed coding and informed embedding to design a robust, high capacity watermark. IEEE Trans. Image Process. **13**(6), 792–807 (2004)

19. B. Oztan, G. Sharma, Multiplexed clustered-dot Halftone watermarks using bi-directional phase modulation and detection, in *Proceedings of 2010 IEEE 17th International Conference on Image Processing* (2010)

20. A. Pramila, A. Keskinarkaus, T. Seppänen, Multiple domain watermarking for print-scan and JPEG resilient data hiding, in *Proceedings of the 6th International Workshop on Digital Watermarking* (2007)

21. W.L. Root, An introduction to the theory of the detection of signal in noise. Proc. IEEE **58**(5), 610–623 (1970)

22. K. Solanki, U. Madhow, B.S. Manjunath, S. Chandrasekaran, I. El-Khalil, Print and scan resilient data hiding in images. IEEE Trans. Inform. Forensics Secur. **1**(4), 464–478 (2006)

23. A. Trémeau, D. Muselet, Recent trends in color image watermarking. J. Imaging Sci. Technol. **53**(1), 10201-1–10201-15 (2009)

24. A. Varna, S. Rane, A. Vetro, *Data Hiding in Hard-Copy Text Documents Robust to Print, Scan, and Photocopy Operations* (Mitsubishi Electric Research Laboratories, Cambridge, 2009)

25. R. Villan, S. Voloshynovskiy, O. Koval, J. Vila, E. Topak, F. Deguillaume, Y. Rytsar, T. Pun, Text data-hiding for digital and printed documents: theoretical and practical considerations, in *Proceedings of the SPIE* (2006), pp. 15–19

26. Z. Wang, A.C. Bovik, H.R. Sheikh, E.P. Simoncelli, Image quality assessment: from error visibility to structural similarity. IEEE Trans. Image Process. **13**(4), 600–612 (2004)

Chapter 3
Text Watermarking

The previous chapter reviewed a number of alternatives for hardcopy communications using printed images as hosts while addressing distortions caused by the printing and scanning processes. An important class of media that is closely related to hardcopy communication is that of text documents. While in natural images there is a rich grey scale or even color content suitable to be modified, in text one usually does not benefit from such a highly diversified host signal. The problem becomes even more challenging when we consider that the watermarked document is to be printed and *remain* watermarked. In this scenario, printed document watermark detection is usually carried out with the help of a flatbed scanner, to digitize the document and identify a possible watermark. This chapter[1] focuses on text documents, which have distinct characteristics that do not allow the direct application of traditional hardcopy watermarking methods.

This discussion is motivated by the fact that the storage and exchange of important text documents with sensitive or classified information are part of almost any company or organization. Suppose that an institution such as a notary's office, a bank, the police, or any large private or public company, possesses both a paper and a secure electronic copy of a sensitive text document. Examples of such documents are birth certificates, legal notes, classified reports, petitions, and declarations. Very often, these important paper copies of documents are exchanged between people and the institutions. The challenge is to define a reliable method for the authentication of such hardcopy documents, so that these copies are claimed as trustworthy, originated from accurate sources.

As we have discussed in previous chapters—and will further discuss throughout the book—traditional authentication methods in paper form include bar codes, holographic and plastic seals, physical paper watermarks, and authorized personnel handwritten signatures. With these alternatives, however, modifications in the text

---

[1]©2008 IEEE. Some paragraphs in this chapter are reprinted from [13] with permission license no. 4233391357964. Other parts are reprinted from [12] with permission license no. 4233400838109.

can be carried out unnoticeably, changing the meaning of sentences or even of the whole document. Documents with barcodes, for example, can be scanned, modified, and re-printed, and the barcode will still be the same. Handwritten signatures can always be forged, and stamps counterfeited. Additionally, all these strategies cause a visible impact to the original document, which is an aspect often undesired.

The remainder of this chapter will discuss popular methods for effective text authentication. All of these methods can serve as an authenticating mark or simply as an additional side message. Namely, we focus on text modulation techniques, where characteristics of the text are modified according to the message to be transmitted. The modulation can be done via affine geometric transformations or luminance modifications, for example. Independently of the technique employed, the goal is to convey the information in an invisible or "discrete" fashion, such that the readability of the text is not affected.

Before discussing the hardcopy text watermarking alternatives, we revisit the effects of the print and scan operation.

3.1 The Effects of Printing and Scanning

When a document is printed and scanned, the rescanned document may look similar to the original, but it will be distorted during the process. Depending on the distortion, an ineffectively designed embedded watermark may be lost in the process. One may look at the problem from a communications perspective and treat the print and scan process as a communication channel, with a given frequency response, delays, spreading function and noise addition. Actually, when a document is printed, the original work, and consequently the digital watermark embedded, are transmitted over a continuous channel (the paper) and suffer various distortions like low-pass filtering, rotations, scalings, translations, contrast and luminance adjustments, in addition to various types of noises. Considering that watermarks have an energy constraint due to the invisibility requirement, most of these distortions can not be overlooked. Hence, suitable theoretical models for the PS process have been proposed, which will be reviewed in Sect. 3.1.3. These models characterize the problem and provide the basis to the design of efficient schemes. In the following some characteristics inherent to printers and scanners are discussed.

3.1.1 Device Characteristics

3.1.1.1 Scanning

The sampling process of a scanner is mainly affected by the following elements:

- The sampling process of some scanners is slightly different in the vertical direction and in the horizontal direction. Sampling in the vertical direction is normally accomplished with optics only. In contrast, sampling in the horizontal

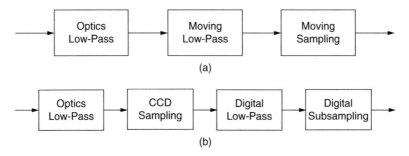

Fig. 3.1 Summary of the scanning procedure. (**a**) Vertical direction. (**b**) Horizontal direction

    direction is achieved by means of an optical sampling and a digital subsampling, often causing non-uniform sampling.

- It is known from the sampling theorem [26] that the sampling rate must be at least twice that of the highest frequency present in the continuous signal to be sampled. In a scanner, a lens-mirror optical system focuses the field of view onto the CCD (Charge Coupled Device) array for acquisition. The optical system is built so that it provides a high-frequency cut that forcefully filters out frequencies higher than half the available horizontal sampling rate.
- A carriage that moves along the vertical direction contains the optical system and the CCD array, and produces samples at the desired resolution. This carriage does not pause its motion when the CCD cells are in acquisition stage and gathering the charges. Hence, the scanned image suffers from blurring and low-pass filtering.

    Figure 3.1a, b show a block diagram of the scanning sampling process in the vertical and horizontal directions, respectively.

3.1.1.2 Printing

There are several printing technologies that are suitable for producing hard copies of images. They include [14] inkjet printing, laser printing, dye sublimation, thermal wax transfer, liquid toner laser transfer, offset printing, and so on. Perhaps the most popular among these are laser and inkjet printers. They offer high quality and high performance at affordable prices. For this reason, focus is given to these two kinds of printers, which work with halftoning technologies.

    Halftoning refers to the process where an image with several levels is converted to a binary black and white image, due to bit-depth limitation of the printing or displaying device. This image, when created by a properly designed halftoning algorithm, looks like a gray scale image if viewed from an appropriate distance, due the low-pass characteristics of the Human Vision System (HVS) [31]. Color images are also printed in halftone [4], but we initially focus the discussion on gray scale images only, as the fundamental theory can be extended to the color case.

    Three popular classes of halftoning algorithms exist: (1) optimization techniques; (2) ordered dithering; and (3) error diffusion. Generally, the optimization techniques present very good perceptual results, but have a high computational complexity in comparison to the other methods. For this reason, although popular, they are not as used as order dithering or error diffusion. Ordered dithering techniques are more commonly applied in laser printers, whereas error diffusion are often used in inkjet printers.

A halftone image has a wide spectrum, and its energy is concentrated at harmonics of the halftone frequency [21]. Different replicas in the spectrum of a scanned halftone overlap, and frequency components may peak with each other to form a periodic visible pattern. This pattern is known in literature as the *Moiré Pattern* [25]. The reader is referred to [14] and [55] for a comprehensive tutorial and additional references on halftoning algorithms.

3.1.2 The Halftoning Process

Many of the hardcopy text watermarking algorithms treat binary images that are eventually to be printed, as gray scale images. In this context, a clear understanding of the halftoning algorithm in the printing process is very important, so that statistical characteristics of the real output of the watermarked signal are known.

As mentioned above, there are two popular types of halftoning algorithms: error diffusion and ordered dithering [42, 53, 56]. The following description is focused on ordered dithering, which, with some variations (depending on the device), is commonly applied in modern laser printers.

Let s be a digital image of size $M' \times N'$ with $L + 1$ levels in the range $[0,1]$ (where **0** represents **white** and **1** represents **black**). A halftone image (binary) b is generated from s, using the ordered dithering halftoning algorithm. The essence of this method lies on the size and on the coefficients of the *dither matrix* D_H (also known as halftone screen) of size $J \times J$, where each coefficient represents a threshold level. Each coefficient in D_H takes a value in the range $\{0, 1/L, 2/L, \ldots, 1\}$. The binary output image b is given by an element-by-element thresholding operation between the pixels in s and the coefficients in D_H.

A block diagram of this operation, with a dither matrix with typical coefficient values created by Bayer in [6], is shown in Fig. 3.2.

In general, the size of the dither matrix is much smaller than the image size, that is, $J \ll M'$ and $J \ll N'$. Therefore, $D_H(m, n)$ is periodically replicated so that $\tilde{D}_H(m, n) = D_H(m \bmod J, n \bmod J)$, where mod is the modulo operation, and the entire image is covered. The input-output relationship of ordered dithering can be mathematically described by:

$$b(m, n) = Q_{H\tilde{D}_H(m,n)}(s(m, n)) \tag{3.1}$$

$$= \begin{cases} 0 & \text{if } s(m, n) < D_H(m \bmod J, n \bmod J) \\ 1 & \text{otherwise} \end{cases} \tag{3.2}$$

The output '0' represents a white pixel (do not print a dot), and '1' represents a black pixel (print a dot). The coefficient values in D_H are approximately uniformly

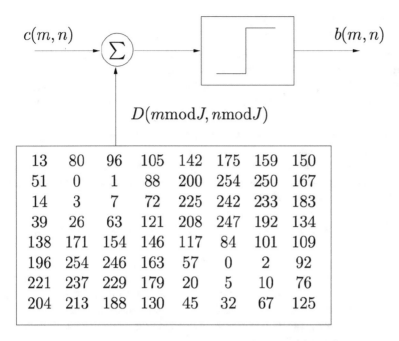

Fig. 3.2 Diagram illustrating pixel-by-pixel operation used in ordered dithering

distributed [56] and clearly have a direct effect on the quality of the halftoned image
b. Commonly used dithering patterns are "clustered-dot," which concatenates black
and white points, and "dispersed-dot," which can be formed from a matrix with low
spatial correlation among the coefficients [53, 56]. Ordered dithering with different
dither matrices is a very popular algorithm in low-cost printing devices and it is
discussed in Sect. 3.1.3.

Many models for the PS operation have been presented in literature, with
variable complexity and distortions considered. This section describe an analytical
PS channel model that includes several characteristics that influence the detection
performance of the text watermarking systems discussed and still allows a mathematical tractability in the analyses. We include a detailed description of the physical
process in an attempt to justify the noise models employed. Depending on the class
of algorithm considered, determining this noisy channel model can be an essential
task in the design of effective hardcopy watermarking algorithms. It is important to
notice that the channel model discussed is also valid for any 2D bar code [21, 57, 61].

3.1.3 Print and Scan Channel Analytical Models

The PS process is often modeled as a noisy communications channel. The distortions induced by this channel are commonly divided in two parts: pixel value
distortions and geometric distortion.

3.1.3.1 Pixel Value Distortions

A model for the luminance alterations suffered by the pixels of a printed and scanned image is presented next. Only grey scale images are considered, and an extensive description and references about color variations is available in the literature [50].

Analytical models of the PS channel have been presented in the literature [21, 52, 57], and they assume that the process can be generally modeled by low-pass filtering, the addition of Gaussian noise, and non-linear gains, such as brightness and gamma alterations.

In the model proposed by Degara-Quintela and Pérez-González [21], for example, the luminance value for a pixel with coordinates (m, n) that has passed through the PS channel is described by

$$y(m, n) = g[s(m, n) \circledast h(m, n)] + f[g(s(m, n))]\eta_1(m, n) + \eta_2(m, n) \qquad (3.3)$$

where s and y are respectively the original and the printed and scanned images, and η_1, η_2 are assumed zero-mean, mutually independent, white Gaussian noise. The operator $\circledast$ represents convolution, h is a low-pass filter, and $g(\cdot)$ and $f(\cdot)$ are generally non-linear effects which represent the gain characteristics of the devices. This model performs satisfactorily in many applications, such as multi-level 2D bar codes, as discussed in [21, 57, 61].

The model in (3.3) provide as simple representation of the pixel value distortion. However, it does not consider some of the physical effects included in the process, such as the halftoning operation. Specifically, by investigating the noise effects induced by the halftoning algorithm, it becomes clear that there exists a dependence between the input luminance of the algorithm and the statistical moments in the output. This characteristic is exploited in the detection metrics of some text watermarking algorithms, such as TLM. Hence, a modified analytical model which accounts for the halftone noise is proposed, followed by modifications of the function $g(\cdot)$.

Therefore, the printing model is described by:

$$y_{pr}(u, v) = g_{pr}[b(u, v)] \circledast h_p(u, v) + \eta_1(u, v) \qquad (3.4)$$

where y_{pr} is the continuous printed image formed from the discrete image s. The continuous coordinate system (u, v) is used to represent the digital signals in the analog domain, that is, $b(m, n) \rightarrow$ D/A $\rightarrow b(u, v)$, where $b(m, n) = Q_H[s(m, n)]$ is the halftoned version (before printing) of $s(m, n)$. The term h_p represents a low-pass effect due to toner or ink spread [45] and η_1 represents the microscopic ink and paper imperfections.

The scanning process is formed by the analog optical acquisition followed by sampling and quantization to digitize the image. Notice that $\sin(x)/x$ and aliasing effects due to the digital to analog (D/A) and analog to digital (A/D) conversions are disregarded. The image to be scanned is illuminated and the reflected intensity is converted into an electrical signal and captured by a CCD sensor.

The analog image acquisition process of the scanning operation can be described by:

$$y(u, v) = g_s\{y_{pr}(u, v) \circledast h_s(u, v)\} + \eta_3(u, v)$$

$$= g_s\left\{\{g_{pr}[b(u, v)] \circledast h_p(u, v) + \eta_1(u, v)\} \circledast h_s(u, v)\right\} + \eta_3(u, v),$$

(3.5)

where y is the printed and scanned signal and h_s is a low-pass filter for modeling the optics and the motion blur caused by the interactions between adjacent CCD arrays elements [21]. η_3 is a noise term which models the illumination noise and microscopic particles in the scanner surface.

The linear systems h_p and h_s in (3.5) model the point-spread functions of the printer and of the scanner, respectively. Considering that the low-pass effect due to scanning prevails over that due to printing, i.e. $h_s \circledast h_p \approx h_s = h$, (3.5) can be written as

$$y(u, v) = g_s\left\{\{g_{pr}[b(u, v)] + \eta_1(u, v)\} \circledast h(u, v)\right\} + \eta_3(u, v)$$

(3.6)

Because the scanning operation involves sampling and quantization, the final scanned image is represented in the digital domain, that is, $y(u, v) \rightarrow$ A/D $\rightarrow$ $y(m', n')$, where A/D represents analog to digital conversion. To express the input-output relationship of the PS system, the digital coordinate system (m', n') is approximated to the original (m, n) coordinates, although the coordinate system (m', n') usually does not correspond exactly to the original (m, n) coordinates. This difference does not affect significantly the input-output relationship of the PS system if the same resolution is used for both printing and scanning operations. When this is not the case, it is possible to include a scaling term to the model, as discussed in Sect. 3.1.3.3.

The digital scanned image is written as

$$y(m, n) = g_s\left\{\{g_{pr}[b(m, n)] + \eta_1(m, n)\} \circledast h(m, n)\right\} + \eta_3(m, n),$$

(3.7)

where η_3 now combines the analog distortions represented by η_3 in (3.5), the CCD electronic noise and the quantization noise due to the A/D operation.

The terms in (3.7) are further discussed in the following.

3.1.3.2 Gains $g_{pr}(\cdot)$ and $g_s(\cdot)$

In practice, when toner or black ink particles are applied over the paper, they do not present a null reflectance, causing a luminance gain to the printed image [42]. This distortion is modeled by:

$$g_{pr}(m, n) = b(m, n)\alpha(m, n),$$

(3.8)

where α is a gain and depends on the device being used. The parameter $\alpha(m, n)$ is modeled as constant for a small region (an area corresponding to one fifth of a page, for example), but it does vary slightly throughout a full page due to non-constant printer toner distribution.

The response of scanners also vary depending on the device, and scanners usually present a non-linear gain represented by

$$g_s(m, n) = [x(m, n)]^\gamma.   \qquad (3.9)$$

where $x(m, n) = \{g_{pr}[b(m, n)] + \eta_1(m, n)\} \circledast h(m, n)$.

Notice that some authors [21, 52, 60] describe the PS gain $g(m, n)$ as

$$g(m, n) = \beta(m, n)[x(m, n)]^\gamma.   \qquad (3.10)$$

In contrast to these models, it is also possible to describe the printing gain g_{pr} and the scanning gain g_s separately, as done in this section. Although a $\beta(m, n)$ term is not included in (3.9), this effect is replaced by the gain α in $g_{pr} = \alpha(m, n)b(m, n)$ in the model discussed.

Actually, there are many sources of noise that are simplified and concise in the model of Eq. (3.7). The noise power is slightly stronger in the moving direction of the carriage in scanner, due to the stepped motion jitter, that causes random sub-pixel drift. This indicates that η is not perfectly symmetric in both directions. Significant voltage may go through the scanner at every clock pulse, and can cause interference on the analog signal of the scanner. Fluctuations in the intensity of light also alter the acquisition, as well as physical elements such as printer spots, type of paper used, and dust and scratches in the scanner [21].

3.1.3.3 Geometric Distortions

A very common procedure when scanning a paper document is as follows. Initially, the user places an image or text to be scanned on a flatbed scanner. A low resolution preview of the image is shown, and the user manually selects, with the help of a graphic user interface (GUI), a cropping region to be scanned at high resolution. In this process the final discrete image suffers three kinds of distortions: rotation, scaling, and cropping.

1. **Rotation:** when a picture is placed on the flatbed, a small misalignment is usually present, producing a rotation distortion. It has been observed [9, 35, 52] that this rotation is usually less than $3°$. Therefore, the rotation process is described by the simple equation:

$$y(m, n) = c(m \cos \phi - n \sin \phi, m \sin \phi + n \sin \phi), \qquad |\phi| \leq 3°   \qquad (3.11)$$

Although not very perceptually significant, this rotation can be a major source of interference in the detection process of a watermarking system.

2. **Scaling:** another distortion is often caused by a different resolution used by the printer and by the scanner, resulting in an image larger or smaller than the original. Therefore, a certain amount of re-scaling of the document occurs depending on the chosen resolutions. This distortion can be represented by a scaling factor λ_s in the spatial coordinates in the input-output relationship of the PS process [35] such that

$$y(m, n) = s(\lambda_s m, \lambda_s n), \tag{3.12}$$

where y is a printed and scanned image and s is the original image. To simplify the notation, however, λ_s is assumed 1 and it is omitted from the PS model discussed.

3. **Cropping:** finally, prior to the high resolution scanning, the scanned image becomes either manually cropped or zero-padded (due to the black background), which are generally referred to as a cropping distortion.

A substantial mathematical analysis of the rotation, scale, and cropping (RSC) geometric distortion induced by PS is given in [34].

As observed in the conducted experiments, TLM is naturally robust to small amounts of rotation distortions as well as robust to scaling. Moreover, cropping will not be an issue provided that no character or symbol (required for the detection) is cropped during the manual selection of the area to be scanned. If it is assumed that cropping may occur, the use of error correcting codes need to be employed to reduce the errors caused by this type of distortion. Affine and cropping distortions are not addressed in this work, and the reader is referred to [20] for a reference concerning error correcting codes for cropping.

3.1.3.4 Effects Induced by the Halftone

As describe in Sect. 3.1.2, the halftoning algorithm quantizes the input to 'print a dot' and 'do not print a dot' according to the dither matrix coefficients. This quantization causes a predictable effect on the variance, skewness, and kurtosis of a halftoned region, according to the input luminance. This characteristic, which is used by some text watermarking algorithms for improved detection, is demonstrated analytically in the following.

Variance

The variance of a halftone block b_0 of size $J \times J$ is given by:

$$\sigma_{b_0}^2 = \frac{1}{J^2} \sum_{m=1}^{J} \sum_{n=1}^{J} [b_0(m, n) - \bar{b}_0]^2 \tag{3.13}$$

$$= \frac{1}{J^2} \sum_{m=1}^{J} \sum_{n=1}^{J} b_0^2(m, n) - 2b_0(m, n)\bar{b}_0 + \bar{b}_0^2, \tag{3.14}$$

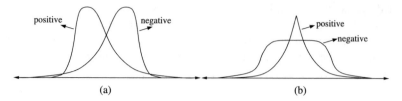

Fig. 3.3 Illustration of the effect of positive and negative (**a**) skewness and (**b**) kurtosis, ©2009 IEEE. Reprinted from [12] with permission license no. 4221170750957

where $b_0(m, n) \in \{0, 1\}$, $\bar{b}_0 = (1/J^2) \sum_{m=1}^{J} \sum_{n=1}^{J} b_0(m, n)$ and J^2 is the number of coefficients in the dithering matrix D_H. Since $b_0(m, n) \in \{0, 1\}$, $b_0^2(m, n) = b_0(m, n)$, and (3.14) can be written as

$$\begin{aligned} \sigma_{b_0}^2 &= \bar{b}_0 - 2\bar{b}_0\bar{b}_0 + \bar{b}_0^{\,2} \\ &= \bar{b}_0 - \bar{b}_0^{\,2}. \end{aligned} \tag{3.15}$$

Similar analysis are performed for the skewness and kurtosis, as shown in the following.

Skewness

The skewness measures the degree of asymmetry of a distribution around its mean [41]. It is zero when the distribution is symmetric, positive if the distribution shape is more spread to the right and negative if it is more spread to the left, as illustrated in Fig. 3.3a.

The skewness of a halftone block b_0 of size $J \times J$ is given by:

$$\gamma_{1b_0} = \frac{\dfrac{1}{J^2} \sum\limits_{m=1}^{J} \sum\limits_{n=1}^{J} [b_0(m, n) - \bar{b}_0]^3}{\sigma_{b_0}^3} = \frac{\bar{b}_0 - 3\bar{b}_0^{\,2} + 2\bar{b}_0^{\,3}}{\left(\bar{b}_0 - \bar{b}_0^{\,2}\right)^{3/2}} \tag{3.16}$$

Kurtosis

The kurtosis is a measure of the relative flatness or peakedness of a distribution about its mean, with respect to a normal distribution [41]. A high kurtosis distribution has a sharper peak and fatter tails, while a low kurtosis distribution has a more rounded peak with wider "shoulders," as illustrated in Fig. 3.3b.

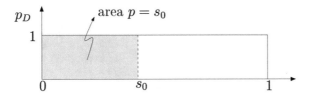

Fig. 3.4 Uniform distribution p_D assumed for the coefficients of the dithering matrix, ©2009 IEEE. Reprinted from [12] with permission license no. 4221170750957

The kurtosis of a halftone block b_0 of size $J \times J$ is given by:

$$\gamma_{2b_0} = \frac{\frac{1}{J^2} \sum_{m=1}^{J} \sum_{n=1}^{J} [b_0(m,n) - \bar{b}_0]^4}{\sigma_{b_0}^4} - 3$$

$$= \frac{\bar{b}_0 - 4\bar{b}_0^2 + 6\bar{b}_0^3 - 3\bar{b}_0^4}{\left(\bar{b}_0 - \bar{b}_0^2\right)^2} - 3 \tag{3.17}$$

To derive $\sigma_{b_0}^2$, γ_{1b_0} and γ_{2b_0} as a function of the input luminance $s(m,n)$, $b_0(m,n)$ must be generated from a constant grey level region, that is, $s(m,n) = s_0$, $m,n = 1, \ldots J$, where s_0 is a constant. Assuming that the dithering matrix is approximately uniformly distributed as illustrated in Fig. 3.4, the probability p of $b(m,n) = 1$, which is $\Pr[s_0 > D_H(m,n)]$, is given by

$$p = \Pr[s_0 > D_H(m,n)] = \frac{1}{J^2} \sum_{b(m,n)=1} b(m,n)$$

$$= \frac{1}{J^2} \sum_{m=1}^{J} \sum_{n=1}^{J} b(m,n) = \bar{b} \approx s_0 \tag{3.18}$$

as illustrated in Fig. 3.4. Substituting this result into (3.15)–(3.17), yields

$$\sigma_{b_0}^2(s_0) = s_0 - s_0^2 \tag{3.19}$$

$$\gamma_{1b_0}(s_0) = \frac{s_0 - 3s_0^2 + 2s_0^3}{\left(s_0 - s_0^2\right)^{3/2}} \tag{3.20}$$

$$\gamma_{2b_0}(s_0) = \frac{s_0 - 4s_0^2 + 6s_0^3 - 3s_0^4}{\left(s_0 - s_0^2\right)^2} - 3 \tag{3.21}$$

where $\sigma_b^2(s_0)$, $\gamma_{1b}(s_0)$ and $\gamma_{2b}(s_0)$ represent respectively the variance, the skewness, and the kurtosis of a halftoned block that represents a region of constant luminance s_0.

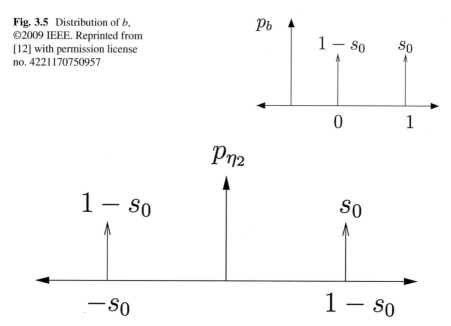

The halftone quantization function Q_H that generates b is binary and without loss
of generality, it is possible to decompose b into a constant term $\bar{b}$ added to a noise
term η_2:

$$b(m,n) = \bar{b} + \eta_2(m,n), \tag{3.22}$$

The noise η_2 has its variance given by $\sigma_{\eta_2}^2 = \sigma_b^2(s_0)$, and it is distributed according
to $\eta_2 \in \{-s_0, 1 - s_0\}$, as illustrated in Fig. 3.6.

Using the model in (3.22) and the fact that $\bar{b} \approx s_0 = s(m,n)$ within a constant
grey level region, (3.7) can be modified to express the input-output relationship of
the PS process as

$$y(m,n) = g_s\left\{\{g_{pr}\{s(m,n) + \eta_2(m,n)\} + \eta_1(m,n)\} \circledast h(m,n)\right\} + \eta_3(m,n), \tag{3.23}$$

assuming that $s(m,n)$ is constant in a small region corresponding to the size of a
halftone block.

The assumption that the dither matrix is approximately uniformly distributed
indicates that the ratio (number of black pixels)/(total number of pixels) increases
linearly according to the input value in the halftone algorithm. When this is assumed,
the result of the analysis of the halftone variance is valid for different dither
matrices (clustered-dot [55], dispersed-dot [55], etc.). Notice, however, that other
characteristics such as the spectral distribution do vary according to dither matrix
employed [55].

Comments on σ_b^2, γ_{1b} and γ_{2b}

The halftone signal b_0 is binary and it is distributed according to $b_0(m, n) \in \{0, 1\}$, with probabilities $1 - s_0$ and s_0, respectively, as illustrated by p_b in Fig. 3.4. Because the variance, the skewness and the kurtosis of b_0 depend on s_0, these moments can be used as detection metrics in text luminance modulation and multi-level bar codes.

The use of the variance is justified because the largest variance due to halftoning occurs in the middle of the input range ($s_0 = 0.5$), where the black and white areas are approximately equal, with the greatest dispersion. This characteristic is observed in [56], where perceptual characteristics of halftone are investigated.

Regarding the skewness, it is equal to zero when $s_0 = 0.5$ and the distribution of b_0 is symmetric, represented by two peaks of equal probability. The two symmetric peaks also flatten the distribution of y in (3.7), minimizing the kurtosis. When $s_0 < 0.5$, b is composed of more white dots than black dots, leaning the distribution of y to the left and causing a positive skewness. The opposite occurs when $s_0 > 0.5$, yielding a negative skewness. Likewise, the distribution of y becomes more peaky as s_0 approaches the limits of the luminance range, consequently increasing the kurtosis.

An example illustrating the effect of a halftone variance level that depends on the input luminance is given in Fig. 3.7a, where two curves are presented. The black curve ('Theoretical') represents the theoretical variance determined in (3.15). The grey curve ('Bayer') represents the variance of a halftone block (before PS) generated using the Bayer dithering matrix [6]. Figure 3.7d illustrates the output image of the halftone algorithm (before printing) using the Bayer dither matrix [6] for input values $0, 15/L, 30/L, \ldots, 1$.

Similar experiments are presented regarding the skewness and the kurtosis, as shown in Fig. 3.7b and c, respectively. These figures illustrate that the analyses of Sect. 3.1.3.4 are in accordance with the results obtained from a practical halftone matrix.

3.1.3.5 PS Channel Conclusions

The PS models discussed above serve as a basis for the design and evaluation of the hardcopy watermarking systems such as those described in the remainder of this chapter. We paid particular attention to the model that includes a term representing the halftone signal. Specifically, by explaining the noise effects induced by the halftoning algorithm, it has been shown that higher order statistical moments such as the variance, the skewness, and the kurtosis of a region vary according to the average luminance of that region. This is exploited by TLM algorithms, where this dependence among the input luminance and the statistical moments in the output also occurs after the PS process. This yields the usage of improved detection metrics to the TLM hardcopy watermarking systems.

Now that the PS channel has been discussed, in the following we describe popular text hardcopy methods.

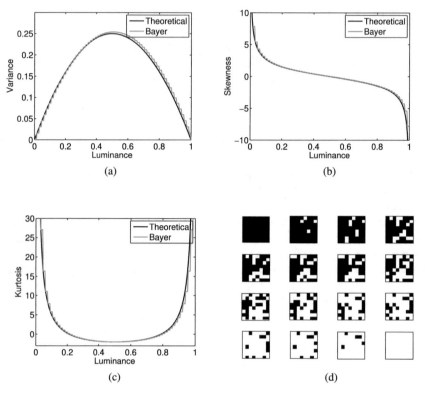

Fig. 3.7 Illustration of how the luminance level input into a halftone algorithm affects the statistics of the output, ©2009 IEEE. Reprinted from [12] with permission license no. 4221170750957. (**a**) The effect of variance dependent on the input luminance. (**b**) The effect of skewness dependent on the input luminance. (**c**) The effect of kurtosis dependent on the input luminance. (**d**) Illustration of halftone blocks

3.2   Text Watermarking via Affine Transformations

Among other types of media, text documents show very characteristic properties: binary nature, distinct block/line/word patterning, and a clear separation between foreground and background areas. Hence, algorithms specific to text are required to deal with these issues. Unfortunately, the techniques presented in the previous chapter (Hardcopy Image Watermarking) that make use of linear transforms present an apparent drawback when used to watermark text documents. In general, any modification in the transform representation of a text image is spread over the whole document, including the white background of a text document. Therefore, noise-like patterns become visible in a constant intensity region such as white paper, violating the fundamental perceptual transparency requirement of watermarking systems.

Following an extensive survey on text watermarking, and in agreement with [32], it has revealed that a rather small number of researches have been developed in this area, in comparison to image, audio and video watermarking. However, an increasing popularity of applications as digital library and digital notary's shall prove text watermarking as an essential element for document authentication and copyright protection in the years to come.

An efficient way to convey information over text documents is perform small geometric affine transformations on text elements. In this case, we refer to text elements as single characters, words, lines or even whole paragraphs.

A landmark paper and a very good reference on text watermarking that deals with geometric transformation has been published by Brassil et al. in [16]. In their work, the authors describe and compare several mechanisms to watermark documents and several other mechanisms for decoding the marks after documents have been subjected to common types of distortion. Their idea exploits the fact that each copy of a text document can be made imperceptibly different by repositioning or modifying the appearance of different elements of text, i.e., line, words, or characters. Their discussion is mainly focused on the word-shift and line-shift algorithms, based on [15, 38, 40]. The line-shift algorithm alters the position of a line by moving it upward or downward depending on the binary signal to be embedded. The same idea applies for word-shifting, where a word is moved left or right according to the embedded message.

3.2.1 Line-Shift Coding

Using this technique, a watermark is embedded in a text document by vertically displacing an entire text line, as illustrated in Fig. 3.8.[2] There are several ways to perform the encoding process, and a typical implementation is to move a line up or down, while the line immediately above or below (or both) is left unchanged.

> How many roads must a man walk down before you
>
> can call him a man? Yes, and how many seas must a
>
> white dove sail, before she sleeps in the sand?
>
> ───
>
> How many roads must a man walk down before you
>
> **can call him a man? Yes, and how many seas must a**
>
> white dove sail, before she sleeps in the sand?

Fig. 3.8 Illustration of line-shift coding. The middle line in the upper paragraph is slightly, yet unnoticeably, moved up. In the lower paragraph, the displacement is visible with the superposition of both the original and the shifted middle line

---

[2]Text sample in this figure extracted from the lyrics of 'Blowing in the Wind', Bob Dylan, 1963.

According to [15], an important feature of this marking technique is the blind decoding process, as it is not necessary to have a copy of the original document (not marked) for comparison during decoding. This is possible because of the generally uniform interline space characteristic of a text document within a paragraph. In this manner, the presence or absence of a watermark can be detected only by the inspection of the interline space in the received content. Experiments suggest [16] that vertical line displacements for 1/300 in and less are unnoticed by readers.

Using line-shift coding, a codeword based on an alphabet $\{-1, +1, 0\}$ is created, which means to shift the line up, down or do not shift it. The latter option, 'do not shift', brings more robustness to the watermark, working as a spatial reference. The length of the embedded message is related to the number of lines in each page of the document. Considering that a text document page has about 19 lines, with this coding strategy one can embed $2^{19} = 524{,}288$ different messages.

To locate the lines one can use a projection profile [16], $\varrho(n)$, given by

$$\varrho(n) = \sum_{m=0}^{M} c(m, n), n = t, t+1, \ldots, p \tag{3.24}$$

where $c(m, n)$ represent an image with elements $m = 0, 1, \ldots, M$ and $n = 0, 1, \ldots, N$. The indexes t and p correspond to the top and bottom vertical limits of a subimage containing a single text line, that is,

$$c(m, n) \qquad m = 0, 1, \ldots, M, \quad n = t, t+1, \ldots, p \tag{3.25}$$

Similarly, the vertical profile is given by

$$v(m) = \sum_{n=t}^{p} c(m, n), m = 0, 1, \ldots, M \tag{3.26}$$

The horizontal profile $\varrho(n)$ is a plot of the summation of ON-valued pixels along each row. For a document whose text lines span horizontally, this profile has peaks whose widths are equal to the character height and valleys whose widths are equal to the white space between adjacent text lines. The distance between profile peaks are the interline spaces. This is illustrated in Fig. 3.9.[3]

Due to the visual nature of the profile, up and down line shifts are referred to as left and right line shifts, respectively. An observed profile $\varsigma(n)$ is defined as

$$\varsigma(n) = \varrho(n) + R(n) \tag{3.27}$$

where $R(n)$ are independently identically distributed (i.i.d.) zero-mean Gaussian random variables with variance σ_R^2, and accounts reasonably for a displacement noise, as discussed in [16].

---

[3]Figure extracted from [16].

Fig. 3.9 Profile of a received marked document page

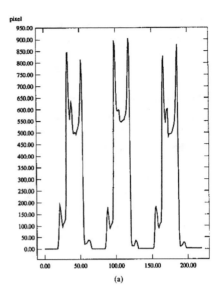

(a)

After receiving the document page, the detector performs a measure of the distance between the baselines of adjacent line profiles (the black dots in Fig. 3.9). Now, based on the known codeword, this measure is used to determine the presence or absence of a mark. Because in a regular unmarked document the lines are uniformly spaced, if the detector verifies line spacing shifts and this matches with the codeword, the presence of the mark is identified.

A correlation detector is also applicable when distortions such as expansions and translations are accurately estimated and compensated for. Given an observed profile $\varsigma(n)$, a line is shifted left if

$$\sum_{p}^{t} \varrho(n)(\varsigma(n-\tau) - \varsigma(n+\tau)) \geq 0 \tag{3.28}$$

Otherwise, it is decided that the line is shift right. In Eq. (3.28), τ is a shift much smaller than the inter-line spacing. Also, it is assumed that the profile $\varrho'(n)$ is in between two unmoved lines, or control blocks. So the three lines unmoved / shifted / unmoved are referred to as blocks 1, 2 and 3, respectively.

A more effective detection alternative when the distortions effects cannot be accurately compensated for, is the centroid detection [40]. Based on the original profile $\varrho(n)$, the centroid of blocks i are

$$r_i = \frac{\sum_{p_i}^{t_i} n\varrho(n)}{\sum_{p_i}^{t_i} \varrho(n)}, \quad i = 1, 2, 3. \tag{3.29}$$

The displacement of the centroids on the observed profile $\varsigma(n)$ are modeled by the addition of $R(n)$, so the control blocks 1 and 3 have centroids

$$U_1 = r_1 + V_1 \quad \text{and} \quad U_3 = r_3 + V_3. \tag{3.30}$$

where V_i are random variable representing the distortion caused by the addition of $R(n)$. Since middle block 2 has been shifted by $\tau > 0$, its centroid is $U_2 = r_2 + V_2 - \tau$ if it is left shifted and $U_2 = r_2 + V_2 + \tau$ if it is right shifted. It has been shown [40] that the variance of V_i is expressed by:

$$\sigma_{V_i}^2 = \frac{\sigma_R^2 f_i}{H_i^2} (\delta_i^2 + (f_i^2 - 1)/12) \tag{3.31}$$

where

$$H_i = \sum_{p_i}^{t_i} \varrho(n) \tag{3.32}$$

$$f_i = t_i - p_i + 1$$

$$\delta_i = r_i - \frac{t_i + p_i}{2}$$

So, the variance $\sigma_{V_i}^2$ is not only dependent of the profile noise variance σ_R^2 but also dependent on the original unmarked profile $\varrho(y)$ through H_i, f_i, and δ_i.

To compensate the effect of translation, Low et al. [40] base their centroid detection on the distance $U_i - U_{i-1}$ between adjacent centroids, instead of the absolute position U_2 of the middle centroid. The decision variable is given by the differences

$$\Gamma_l := (U_2 - U_1) - (r_2 - r_1) \tag{3.33}$$

$$\Gamma_r := (U_3 - U_2) - (r_3 - r_2)$$

of the received centroid separations and the original centroid separations. Γ_l and Γ_r are the distances from middle block 2 to left control block 1 and right control block 3, respectively. In a noise free situation, $\Gamma_l = -\tau$ and $\Gamma_l = \tau$ if the middle block is left shifted, and $\Gamma_l = \tau$ and $\Gamma_l = -\tau$ if it is right shifted. The centroid detection follows the decision rule:

IF $\quad \dfrac{\Gamma_l}{\sigma_{V_1}^2} \leq \dfrac{\Gamma_r}{\sigma_{V_3}^2}, \quad$ assume that middle block 2 is shifted left;

OTHERWISE, assume that middle block 2 is shifted right.

Centroid detection, compared with baseline, is more reliable in the presence of noise, but it needs the original centroids during detection and does not work satisfactorily with word-shift coding [16]. On the other hand, the main advantage of baseline detection is that it performs blind detection, but it has a relatively poor performance with noisy documents [15]. An interesting theoretical analysis of channel capacity using centroid detection with line-shift coding is presented in [39], where the authors show that to achieve maximum capacity the shifts should be normally distributed.

Experimental analyses show that this line-shift coding can survive to the process of print-and-scan, even after several copies [16]. On the other hand, this method also permits anyone to be able to read the embedded information, which is sometimes regarded as a disadvantage for the system.

3.2.2 Word-Shift Coding

The main idea behind this method is altering a text document by horizontally shifting the position of words within text lines to encode a given message. Several text justification rules exist, and a variable spacing in text documents is commonly used to distribute white space when justifying text. This method is least visible when applied to documents with variable spacing, and that, in general, horizontal word displacements of 1/150 in and less go unnoticed [16]. A text sample with that level of displacement is illustrated in Fig. 3.10. Because of the variability of word spacing, the original document is required to perform the detection of the mark. In Fig. 3.10, an example of word-shift coding is given.

A good strategy to encode a text with word-shift coding is to find the largest and smallest spacing between words for each text line. In order to keep the distribution of the words in the line, one should decrement the largest spacing by some amount and increment the smallest spacing by the same amount. This procedure performs the encoding process, which by reallocating words spacing creates a codebook that is used by the decoder to identify the original document recipient.

At the decoder, the locations of text elements are extracted from the received document (it could be scanned from a printed version). These locations are then compared with those from the original documents distributed. The decoder matches one of the original documents and makes use of the codebook to verify the identity of the original document recipient. The detection techniques discussed in Sect. 3.2.1 are generally applied in word-shift coding as well, depending on the level of noise present [16].

Make sure that the fortune that you seek is the fortune that you need

Make **sure** that the **fortune** that you **seek** is the fortune that **you** need

Fig. 3.10 Example of word-shift coding. In the upper phrase, the words *sure, fortune, seek* and *you*, where slightly shifted. In the lower phrase, the displacement is visible with the superposition of both the original and the shifted phrases

Acredito ser o mais valente, nessa luta do rochedo com o mar.

Acredito ser o mais valente, nessa luta do rochedo com o mar.

Fig. 3.11 Example of character-shift coding

A different word-shifting algorithm was developed by Huang et al. in [29]. There, the inter-word spaces are modified such that the spaces represent a sine wave. Attractive advantages of this technique are: the sine wave varies gradually so that local variations may be unnoticed; the message to be embedded is encoded in the phase, amplitude, and frequency of the sine wave(s); its periodic symmetry make decoding easier and more reliable. In addition, the algorithm can be easily set to perform blind or non-blind detection.

In [32], inter-word spaces are also modified to insert information in text documents. However, instead of modulating spaces according to a sine wave, inter-space statistics such as the mean and the variance are used, so words are shifted right or left depending on the required distribution. The encoding rule does not use individual spaces, but the statistics of a number of spaces.

Others word-shifting algorithms are also available in literature [1, 64], but they are essentially variations of the fundamental idea. An important alternative to word-shift coding is character coding technique, which embeds a watermark by modifying an individual character, as illustrated in Fig. 3.11. Modifications include, for instance, a change to an individual character's height or its relative position, and once again, some characters are left unchanged to facilitate decoding. Character encoding has the advantage of a potentially large embedding capacity, since there is a larger number of characters than lines or words in a page. This property also permits redundancy to be used with an error correction code to improve robustness.

3.2.3 Feature Coding

A class of methods to watermark text documents was presented in the previous section, where high-level features such as spacing and relative positions are altered. Another common way to watermark text is by performing modifications on the actual pixels of characters, such as flipping a pixel from black to white (pixel toggling), and vice versa.

In the work of Wu and Liu [63], an image is partitioned into blocks and a fixed number of bits are embedded in each block by modifying some pixels in that block. A given pixel is altered depending on a "flipping score," which rates the perceptual impact of changing that pixel, determined dynamically by observing smoothness and connectivity. The smoothness is measured by the horizontal, vertical, and diagonal transitions in a local window, and the connectivity is measured by the number of the black and white clusters. Illustration of two pixels that would have a clearly different flipping score is given in Fig. 3.12, where flipping the centre pixel

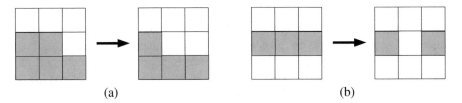

Fig. 3.12 Illustrations of two centre pixels in a 3 × 3 neighborhood that would have different flipping scores

in Fig. 3.12b has a greater perceptual impact than in Fig. 3.12a. To survive the PS channel, the authors suggest the use of visible registration marks to compensate for synchronization distortions.

In [2], the author proposes a feature calibration method for text watermarking, in which two sets of partitions are symmetrically arranged, and the difference between the average features values of each of the two sets is extracted. A text line is divided in two parts, and if the difference between the average width of horizontal strokes exceeds a given threshold, the presence of a watermark is assumed.

Not included in any of the text watermarking classes that have been reviewed so far, the system offered by [17] embeds information into documents by replacing words with their synonyms, not changing the meaning of sentences. The watermark is encoded into the text by rephrasing it minimally, based on a choice of synonyms, word order, and positions of additional blanks for block justification. Although it is not possible to have full access to the algorithm, a clear advantage of this watermark is its capacity to still be embedded in a re-printed version of the document, obtained with the aid of optical character recognition (OCR) techniques, for example. However, it is apparently language dependent, not being directly usable in any language.

3.3 Text Luminance Modulation

In contrast to geometric transformations, the luminance of the characters can also be modified to convey information. This method is usually referred to in the literature as text luminance modulation (TLM). It is acknowledged that the bulk of documents in the office-like class are composed by black text on white background, being referred to as bi-level or binary documents. TLM is able to insert and retrieve information in documents composed by any kind of content, as long as they can be represented by binary text, logos, math symbols, and even line-drawings [24]. In other words, the method can be applied to documents without grey scale or color elements, as is the case in innumerous situations such as office documents, petitions, declarations, scientific articles, and even regular pages in a book. Using TLM, information is embedded by allowing grey level tones to regular text while providing a low perceptual impact. A grey level modulation is added to the original text and it resembles a Pulse Amplitude Modulation (PAM) system. Although these luminance modifications do not affect the perceived text quality, they are easily detected by a scanner, and can be decoded to retrieve the embedded message.

It is acknowledged that the bulk of documents in the class of the examples just mentioned are composed by black text on white background, being referred to as bi-level or binary documents. Using TLM the insertion of a watermark is done by allowing grey level tones to regular text, respecting, of course, the fidelity criteria. So a grey level pattern, or watermark, is added to the black regions of a text image. Some aspects of the algorithm make it intuitively interesting for text documents:

- Due to the high contrast nature of the documents, luminance modifications in the pixel value of a character or symbol are generally hard to perceive.
- From a macro view, a text region itself is a high frequency and textured area, where modifications are not easily noticed.
- Several experiments with varying gains for the embedded watermark were performed, and surprisingly, it was found that gains up to 50 levels of luminance (in the 0–225 scale) go unnoticed in a sample document.
- From the TLM basic idea, any kind of luminance alteration are performed on the 'black' pixels of a document. The convenience, in this case, is that one knows exactly the statistics of the original host signal, that is, totally black. That can facilitate the detection process.
- Most of the office-like documents are composed by black text on white paper, allowing a reasonable application range for TLM.
- A rule that applies not only for this, but for any hardcopy watermarking scheme, is that the digital (pre-printed) version of the watermarked image is not required to have a high visual quality. It is only the *printed* version of the watermarked image that should be visually similar to the printed version of the original image.

Early experiments have illustrated that the above remarks are generally true in practice, motivating further investigation.

3.3.1 General Embedding

Using TLM, information is inserted in a document by altering its luminance through an embedding function $\mathscr{E}(\cdot)$ to insert a watermark w into c, where c is a binary image of size $M' \times N'$ representing a text document. Working in the range $c \in \{0, 1\}$ and $w \in [0, 1]$, where 0 represents white and 1 represents black, the general embedding function is given by:

$$s(m, n) = w(m, n)c(m, n) \tag{3.34}$$

where s is the grey level watermarked version of c, before the PS process.

Notice that in (3.34) the white background is left unchanged and the black regions are modulated by $w(m, n)$. $w(m, n) = 1$ represents no modification and $w(m, n) = 0$ represents the maximum change. The modulating function $w(\cdot)$ is controlled to provide a low perceptual impact, respecting a maximum average distortion given by $E\{[c(m, n) - s(m, n)]^2\} \leq \epsilon_{max}^2$, where ϵ_{max} is a distortion constraint. Practical values for ϵ_{max} are presented in the experiments. Notice that the embedding process described by (3.34) does not rely on a specific function $w(m, n)$. Different data transmission schemes can be applied, and detection can be performed by evaluating amplitude levels, frequency characteristics, among others.

3.3.2 Capacity Upper Bound

In contrast to traditional image watermarking where there is an unknown host luminance, in this case the host function has a constant black luminance value. In the following an upper bound on the theoretical maximum achievable rate of the print and scan process is derived, independently of the code or embedding function $\mathcal{E}$ used for data transmission. This analysis extends generic capacity analyses [57], however considering a limit on the perceptual distortion. It is assumed that in the halftoning process, ordered dithering algorithm is used (common in laser printers). For the upper bound, the use of ideal printer and scanner devices are assumed, and no blurring or synchronization errors are present.

Let r_p and r_s represent the printer and scanner resolution in dots per inch (dpi) and in pixels per inch (ppi), respectively. Consider that the dynamic range $\{0, 1, \ldots, \Lambda\}$ is used, which is simply the set of possible luminance values, usually represented in the $\{0, 1, \ldots, 255\}$ range. Let J be the length in dots of the side of a square halftone dither matrix, or halftone cell. In Fig. 3.2, for example, $J = 8$.

With the above assumptions, the printer outputs $(r_p/J)^2$ halftone cells per square inch. Furthermore, each halftone cell represents up to [57],

$$l = J^2 + 1 \tag{3.35}$$

different grey levels, considering all the possible inputs in the dynamic range $\{0, 1, \ldots, \Lambda\}$. Figure 3.13 illustrates this for $J = 2$, where $2^2 + 1 = 5$ distinct levels are represented. However, limiting the modulation energy to ϵ_{max} due to the perceptual transparency requirement, the available dynamic range is decreased to $\{0, 1, \ldots, \epsilon_{max}\}$, and l in (3.35) is decreased to

$$l = (\epsilon_{max}/\Lambda)J^2 + 1, \tag{3.36}$$

considering a generic halftone matrix with uniform spacing between coefficients. Notice that this reduction occurs because, due to the perceptual limit ϵ_{max}, not all J^2 dots can be changed from black to white, but only a fraction of these. This fraction is given by ϵ_{max} out of Λ.

Assuming that every halftone cell is used as a 2D symbol, it is possible to place up to

$$C = (r_p/J)^2 \cdot \log_2 l$$
$$= (r_p/J)^2 \cdot \log_2(\frac{\epsilon_{max}}{\Lambda}J^2 + 1) \tag{3.37}$$

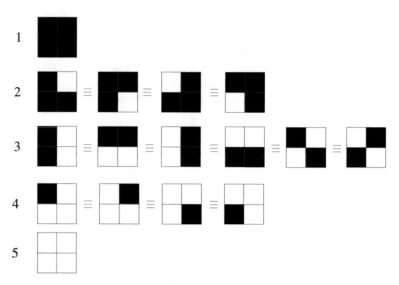

Fig. 3.13 Illustration of $J^2 + 1$ non-equivalent levels in halftoning, for $J = 2$

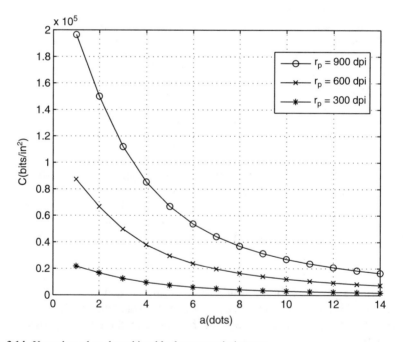

Fig. 3.14 Upper bound on the achievable data transmission rate

bits of information per square inch. A plot of Eq. (3.37) is given in Fig. 3.14. For positive values of J, this upper bound is strictly decreasing, and its maximum is obtained for $J = 1$.

Example

Suppose that the available dynamic range is $\{0, 1, \ldots, 4\}$ and suppose that $r_p = 10$ and $J = 2$, as illustrated in Fig. 3.15. In this figure, $(r_p/J)^2 = 25$ halftone cells per square inch. Supposing that the ordered dithering halftone matrix D_H contains the coefficients given in Fig. 3.16, each halftone cell can assume $J^2 + 1 = 5$ distinct levels. If a limit ϵ_{max} to the dynamic range is not considered, and if each halftone cell is assumed as a 2D symbol, $(r_p/J)^2 \cdot \log_2(J^2 + 1) = 25 \cdot \log_2 5$ bits per square inch can be transmitted in this scenario. However, imposing a limit, say, $\epsilon_{max} = 1$ (dotted line in Fig. 3.16) on the signal energy due to the perceptual transparency requirement, the achievable transmission rate is decreased to

$$C = (r_p/J)^2 \cdot \log_2\left(\frac{\epsilon_{max}}{\Lambda}J^2 + 1\right)$$

$$= (10/2)^2 \cdot \log_2\left(\frac{1}{4}2^2 + 1\right) \tag{3.38}$$

$$= 25,$$

Fig. 3.15 Illustration of halftone cells per square inch for $r_p = 10$ and $J = 2$.

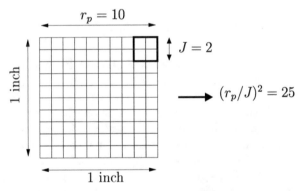

Fig. 3.16 Simple illustration showing how the input luminance affects the output halftone signal

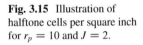

where C is expressed in bits per square inch. Notice that this is a theoretical upper bound only. Theoretically, it is also possible to find a lower bound on capacity by computing the mutual information between input and output of the process. However, in practice, the distortions caused by the PS channel include low pass filtering, quantization, additive noise, and geometric distortions, considerably reducing the embedding capacity.

Considering the constraint ϵ_{max}, these distortions reduce the number of distinguishable levels and consequently reduce the transmission rate. If only the additive Gaussian noise in the channel is considered, it would be possible to derive the reduced capacity using elements of classical information theory. However, this is not the main source of error in the channel, as the low pass filtering and the mis-synchronization due to the geometric distortions have more effect than the noise in the dot-by-dot detection scenario described above. Instead, considering these PS distortions, a practical and robust implementation of TLM has also been considered, which is described in the next section. For this practical implementation a lower bound on the capacity can be expressed on the minimum amount of information transmitted for a given error probability.

3.3.3 Practical Implementation

In a practical implementation of TLM, each element (a character or symbol) in the original digital document is labeled as c_i, with $i = 1, 2, \ldots, K$, where K is the total number of elements. A common standard is to label the elements from left to right, and from top to bottom. An illustration of element labeling is given in Fig. 3.17, where $K = 16$. To embed the watermark, a simple luminance gain is added to individual text elements (characters, symbols, lines, etc.).

Information is embedded by individually altering the luminance of c_i through an embedding function where all pixels in each element have their luminances modulated from black to any value in the real-valued discrete alphabet

$$\Omega = \{\omega_1, \omega_2, \ldots, \omega_S\}$$

of cardinality S, so that each symbol represents $\log_2 S$ bits. Considering a spatial coordinate system for each element, c_i is modulated by a gain w_i, $w_i \in \Omega$. Working in the range $c_i \in \{0, 1\}$ and $w_i \in [0, 1]$, the general embedding function is given by:

$$s_i(m, n) = w_i c_i(m, n), \tag{3.39}$$

where s_i is the modified element. The embedding rate of TLM with this specific implementation is expressed in $\log_2 S$ bits/element. The process is illustrated in Fig. 3.17 for $S = 2$, with a very high gain. In contrast to what is illustrated in this figure, the luminance gain is usually chosen to cause low perceptual impact, such that $E\{[c(m, n) - s(m, n)]^2\} \leq \epsilon_{max}^2$.

Fig. 3.17 Luminance modulation with high gain, ©2009 IEEE. Reprinted from [11] with permission license no. 4221160975332

The printed document contains a hidden, possibly encrypted and channel coded, bit string. The decoder scans the document using resolution r_s and the elements are segmented from the background. Again, each element is labeled as $c_1, c_2, \ldots, c_K$, where K is the number of elements in text. When the paper is properly placed in the flatbed scanner, a small rotation induced by the scanning process does not compromise the labeling process. The detection statistics described in the next section are employed to decode the symbol embedded in each element c_i.

Unlike 2D bar codes [57] or the method based on Eq. (3.37), which have an embedding capacity expressed in bits/in^2, the capacity of the system with this specific implementation depends on the size of the elements, and it is expressed in bits/element.

3.3.4 *Thresholding*

The initial stage of *any* hardcopy text watermarking algorithm is the segmentation process, where the characters from the printed and scanned document are separated from the background by the use of a thresholding function. If not well designed, this can be a major source of error, specially for the single pixel flipping algorithms presented in Sect. 3.2.3. In the TLM case, where the embedded message is transmitted in a element-wise order, a faulty segmentation can also result in an erroneously decoded message. The segmentation quality is affected by character spacing, size, and font style. It is a simpler task to separate from the background the characters in **SIGNAL PROCESSING** than in *SIGNAL PROCESSING*, after both have been printed and scanned. However, when working with usual fonts in office-like documents, for example, experimental analysis have indicated that segmentation is not a critical cause of error in the TLM algorithm, since it does not rely on very small regions or single points to decode a message.

Several text segmentation algorithms have been proposed in literature. Determining and designing segmentation techniques to be specifically applied to TLM is subject of further study. However, a simple global threshold to perform segmentation can work satisfactorily according to the function:

$$s(m,n) = \begin{cases} 0 & \text{if} \quad s(m,n) < T \\ s(m,n) & \text{if} \quad s(m,n) \geq T \end{cases} \tag{3.40}$$

Fig. 3.18 Sample text
scanned at 300 ppi

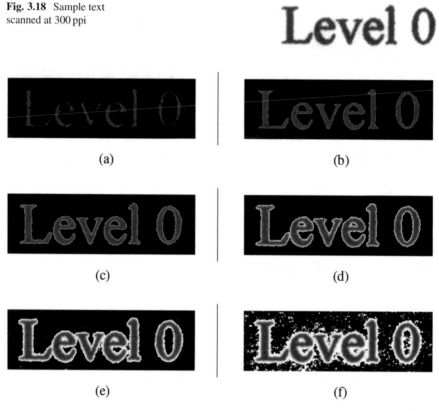

(a)   (b)

(c)   (d)

(e)   (f)

Fig. 3.19 Visual effect of varying the threshold in the text segmentation stage. (**a**) Threshold $T =$ 89. (**b**) Threshold $T = 127$. (**c**) Threshold $T = 178$. (**d**) Threshold $T = 229$. (**e**) Threshold $T = 249$. (**f**) Threshold $T = 254$

The threshold T is manually selected and an adequate value has been determined by experimental analysis. Otsu's segmentation method [46] is an alternative that provides good results [9], choosing the global image threshold that minimizes the intraclass variance of the black and white pixels. However, that method is designed for gray level images, and in the case of text, it leads to results similar to using a simple manually selected threshold. In the following the process of threshold selection and the effects of a faulty segmentation are illustrated.

Consider the string 'Level 0', in Times font type, originally in size 11. The string was printed at 300 dpi, and scanned at 300 ppi, resulting in the image presented in Fig. 3.18. By varying the threshold T on the image (T is given in the 0–255 range), the segmented images presented in Fig. 3.19 are obtained. Notice that the negative image is presented for visualization on paper.

By observing the images in Fig. 3.19, it is possible to note that a small T separates a single character into more than one element, specially in regions formed by thin strokes. On the other hand, a very high threshold unifies separated characters and creates noise elements. In this example, an adequate thresholding value would be $T = 130$, which is the value that often works for typical office papers and laser printers [9].

3.3.5 Attacks

Commonly, the main goal of word, character or line-shift coding algorithms as well as TLM is provide authentication, preventing any tampering with the document. It should be considered that malicious attacks may occur, such that the attacker tries to: (*i*) eliminate the embedded watermark; or (*ii*) modify the embedded watermark.

Consider the following scenario. Suppose the binary representation of the name, place and date of birth of a person is spread over the whole page of a birth certificate, aiming at certifying the document. Considering case (*i*), an attacker does not benefit from tampering with the document such that the watermark is eliminated (as photocopying the document, for example), because the modulation and the information content do not match. Therefore, the document becomes invalid if the watermark is eliminated.

On the other hand, in case (*ii*), the attacker can generate a new document, with a different name and with a new binary modulation pattern over the text. However, because the bit string over the text should be encrypted (as discussed in Sect. 3.3.3) in order to generate this new document the attacker needs to have access to the encryption key used in the embedding process. As with many watermarking schemes [5, 19], it is assumed that a key is known at both the embedding and detection stages. Therefore, the security of TLM relies on the security of the encryption key.

The TLM approach assumes that the encryption key is kept safe, beyond the reach of attackers. For this reason, it is assumed that the document will not be subjected to any attacks. The investigation of scenarios in which the encryption key leaks to attackers is beyond the scope of this book. In the case attacks are assumed, a plausible solution is to use TLM not as the *only* authentication method, but to combine it with traditional authentication techniques such as stamps, physical paper watermarks, handwritten signatures, or plastic seals.

However, even without malicious attacks, watermark detection considering the PS is not a straightforward procedure when a very low error rate is necessary. The next chapter discusses several metrics for efficient watermark detection in the PS channel.

3.3.6 TLM Detection Metrics

Several detection metrica can be used to determine the symbol embedded in an element c_i. Intuitively, the simplest detection metric is the evaluation of the average luminance of the character, as luminance is the modified feature. However, when halftoning algorithms are used in the printing process, detection can be extended by observing that second, third, and fourth order statistical moments of the transmitted symbol also change, depending on the luminance level. In addition, it is shown that an analysis of the spectral characteristics of the received character is also an efficient detection metric.

A thorough analysis of the relationship between the modulated luminance and the higher order statistics of a printed and scanned region is presented in Sect. 3.4. This substantiates the applicability and justifies the use of the proposed metrics. For the first and second moments, an error probability analysis is also presented.

Section 3.4 also discusses combining the detection metrics into a single metric, instead of using them separately. One option [9] is to combine the metrics according to the Bayes classifier, which yields the minimum average classification error for normally distributed patterns. This procedure does not affect the original embedding process and significantly improves the detection performance, as indicated by experimental and theoretical analyses. Using this strategy, it is possible to improve the performance by including other detection metrics not discussed in this book.

3.4 On TLM Metrics

Prior to presenting the detection metrics some comments regarding distortions and statistical assumptions in the PS channel are given in the next section.

3.4.1 Statistical and Distortion Assumptions

By mapping the (m, n) coordinates to a one-dimensional notation, the input-output relationship of the PS process for the i-th character in the document is described by

$$y_i(n) = g_s\left\{\{g_{pr}[s_i(n) + \eta_2(n)] + \eta_1(n)\} \circledast h(n)\right\} + \eta_3(n). \qquad (3.41)$$

Consider the following remarks regarding the PS model in (3.41):

1. The functions $g_{pr}(\cdot)$ and $g_s(\cdot)$ in the above equation represent the printing and the scanning gains, respectively. They are given by $g_{pr}(n) = \alpha(n)s(n)$, $g_s[x(n)] = x^\gamma(n)$. When operating only in a small range of the pixel range $[0,1]$, the parameter γ is close to 1, and $g_s(\cdot)$ can be simplified to a linear model [21]. This is the case in TLM, which usually operates only in the luminance range $[(255 - 40)/L, 1]$ $(L = 255)$, due to the perceptual transparency requirement of the application. Thus, $\gamma = 1$ is assumed in the analyses of this section.
2. Recall that the term $\alpha(n)$ in $g_{pr}(\cdot)$ represents a gain (see (3.8)) that varies slightly throughout a full page due to non-uniform printer toner distribution. Due to its slow rate of change, $\alpha(n)$ is modeled as constant within a given element c_i,

resulting in $\alpha(n) = \alpha_i$, which acts as a gain variable from element to element. Therefore, α_i is assumed constant in n but varying in each realization i, satisfying $\alpha \sim \mathcal{N}(\mu_\alpha^2, \sigma_\alpha^2)$.

3. Moreover, element i is modified by a watermark embedding gain w_i which is constant for the whole element, as described in (3.39), resulting in $s_i(n) = w_i$. Therefore, the term $g_{pr}[s_i(n) + \eta_2(n)]$ in (3.43) is written as $\alpha[w_i + \eta_2(n)]$.

4. Due to the nature of the noise (discussed in Sect. 3.1.3) and based on experimental observations, the noise terms η_1 and η_3 can be generally modeled as zero-mean independent Gaussian noise [21, 57, 61]. Regarding the noise η_2, although it is zero-mean and may be assumed approximately uncorrelated, it is not normally distributed (see Fig. 3.6). However, considering the sum of the several distortions of the channel, it is observed experimentally that the detector output d_{M_i} (defined in (3.42)) can be modeled as normal random variable [51] as supported by the Central Limit theorem, with $d_M \sim \mathcal{N}(\mu_d, \sigma_d^2)$, where d_M, μ_d and σ_d^2 depend on w.

3.4.2 Statistical Detection Metrics

3.4.2.1 Detection by the Sample Mean

The simplest detection metric to determine the symbol embedded in an element c_i is the average luminance of the element, given by (3.42). It is known from detection theory that this detection statistic is the Neyman-Pearson (NP) detector (which minimizes the error probability) when detecting a change in the mean level considering Gaussian noise, which is the framework of the application.

By mapping the (m, n) coordinates to one-dimensional notation, the detection metric d_{Mi} for element i is given by:

$$d_{Mi} = \frac{1}{N_i} \sum_{n=1}^{N_i} y_i(n), \tag{3.42}$$

where N_i is the number of pixels in element i and $y_i(n)$ is the printed and scanned version of $s_i(n)$, according to the PS model described in (3.41). Based on the value of d_{Mi}, it is required to decide on which value of the alphabet Ω element i corresponds to. Due to the noisy PS channel, the detection statistic is a random variable expressed as

$$d_{Mi} = \frac{1}{N_i} \sum_{n=1}^{N_i} g_s \left\{ \{g_{pr}[s_i(n) + \eta_2(n)] + \eta_1(n)\} \circledast h(n) \right\} + \eta_3(n). \tag{3.43}$$

Notice that to find the probability of error, the expected value and the variance of the distribution of d_{M_i} must be determined. For simplicity, in the following a compact notation is used by dropping the index i from d_{M_i} and its derivations.

The expected value of d_{M_i} in (3.43) is

$$\mu_{d_M} = E\{d_M\} = E\left\{ \frac{1}{N} \sum_{n=1}^{N} \{\alpha w + \alpha \eta_2(n) \; \eta_1(n)\} \circledast h(n) + \eta_3(n) \right\}. \tag{3.44}$$

Recalling that $\eta_1(n)$, $\eta_2(n)$ and $\eta_3(n)$ are assumed zero-mean mutually independent random variables and that $cte \circledast h(n) = cte$ given that $\sum_n h(n) = 1$ and cte is a constant, the expected value of d_{M_i} in (3.43) is $\mu_{d_M} = E\{d_M\} = \mu_\alpha w$, where μ_α is the expected value of the gain α.

The detection variance $\sigma_{d_M}^2$ is given by:

$$\sigma_{d_M}^2 = E\{(d_M - \mu_{d_M})^2\}$$

$$= E\left\{ \left(\frac{1}{N} \sum_{n=1}^{N} \{\alpha w + \alpha \eta_2(n) + \eta_1(n)\} \circledast h(n) + \eta_3(n) - \mu_\alpha w \right)^2 \right\}. \tag{3.45}$$

From the mutual independence and zero-mean assumption for $\eta_1(n)$, $\eta_2(n)$ and $\eta_3(n)$, the crossing products $E\{\eta_k(n)\eta_l(n)\}$, $k \neq l$ are disregarded, and (3.45) results:

$$\sigma_{d_M}^2 = \frac{1}{N^2} \sum_{n=1}^{N} \sum_{m=1}^{N} E\{\alpha^2 [\eta_2(n) \circledast h(n)][\eta_2(m) \circledast h(m)]\}$$

$$+ \frac{1}{N^2} \sum_{n=1}^{N} \sum_{m=1}^{N} E\{[\eta_1(n) \circledast h(n)][\eta_1(m) \circledast h(m)]\} \tag{3.46}$$

$$+ \frac{1}{N^2} \sum_{n=1}^{N} \sum_{m=1}^{N} E\{\eta_3(n)\eta_3(m)\} + E\{[w(\alpha - \mu_\alpha)]^2\}.$$

Let $z_1(n) = \eta_1(n) \circledast h(n)$ and $z_2(n) = \eta_2(n) \circledast h(n)$, thus:

$$\sigma_{d_M}^2 = \frac{1}{N^2} \sum_{n=1}^{N} \sum_{m=1}^{N} E\{\alpha^2 z_2(n)z_2(m)\} + \frac{1}{N^2} \sum_{n=1}^{N} \sum_{m=1}^{N} E\{z_1(n)z_1(m)\} + \frac{\sigma_{\eta_3}^2}{N} + w^2 \sigma_\alpha^2$$

$$= \frac{1}{N^2} \sum_{n=1}^{N} \sum_{m=1}^{N} R_{z_2}(m,n)(\sigma_\alpha^2 + \mu_\alpha^2) + \frac{1}{N^2} \sum_{n=1}^{N} \sum_{m=1}^{N} R_{z_1}(m,n) + \frac{\sigma_{\eta_3}^2}{N} + w^2 \sigma_\alpha^2 \tag{3.47}$$

where $R_{z_1}(m,n)$ and $R_{z_2}(m,n)$ are the autocorrelation functions at the blurring filter output, for the input signals η_1 and η_2, respectively. Let $R_{z_1}(m,n) = r_h(m,n) \circledast r_{\eta_1}(m,n)$, by observing the output properties of a linear system with random input [41]. The variables r_{η_1} and r_h represent the autocorrelation functions of η_1 and of the impulse response of h, respectively. Therefore,

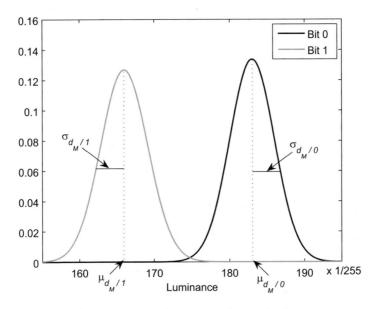

Fig. 3.20 Typical distribution of the detection based on the mean luminance

$$\sigma_{d_M}^2 = \frac{\sigma_\alpha^2 + \mu_\alpha^2}{N^2} \sum_{n=1}^{N} \sum_{m=1}^{N} r_h(m,n) \circledast r_{\eta_2}(m,n)$$

$$+ \frac{1}{N^2} \sum_{n=1}^{N} \sum_{m=1}^{N} r_h(m,n) \circledast r_{\eta_1}(m,n) + \frac{\sigma_{\eta_3}^2}{N} + w^2 \sigma_\alpha^2. \tag{3.48}$$

Since η_1 and η_2 are modeled as uncorrelated noise, $r_{\eta_1}(m,n)$ and $r_{\eta_2}(m,n)$ are represented by an impulse at the origin with amplitude $\sigma_{\eta_1}^2$ and $\sigma_{\eta_2}^2$, respectively. Since $\sum_n h(n) = 1$, $\sum_{m,n} r_h(m,n) = 1$, and (3.48) becomes

$$\sigma_{d_M}^2 = \frac{(\sigma_\alpha^2 + \mu_\alpha^2)\sigma_{\eta_2}^2 + \sigma_{\eta_1}^2 + \sigma_{\eta_3}^2}{N} + w^2 \sigma_\alpha^2, \tag{3.49}$$

where $\sigma_{\eta_2}^2 = (w - w^2)$, as presented in (3.21).

Considering the $S = 2$ (or 1 bit) case, for example, the conditional error probability p_{01} given that bit 1 was transmitted is described by $p_{01} = \Pr(d_M > \lambda_M | \text{bit} = 1)$, where λ_M is a decision threshold. Therefore, $p_{01} = \frac{1}{2}\text{erfc}\left(\frac{\lambda_M - \mu_{d_M/1}}{\sqrt{2\sigma_{d_M/1}^2}}\right)$, where $\mu_{d_M/1}$ and $\sigma_{d_M/1}^2$ are respectively the mean and the variance of d_M for bit 1, as illustrated in Fig. 3.20. Equivalently, if bit 0 is transmitted, the conditional error probability is given by $p_{10} = \frac{1}{2}\text{erfc}\left(\frac{\mu_{d_M/0} - \lambda_M}{\sqrt{2\sigma_{d_M/0}^2}}\right)$, where $\mu_{d_M/0}$ and $\sigma_{d_M/0}^2$ are respectively the mean and the variance of d_M for bit 0, also illustrated in Fig. 3.20.

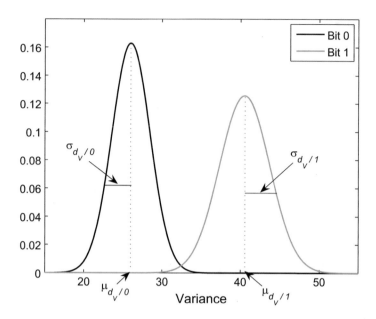

Fig. 3.21 Typical distribution of the detection based on the sample variance

Finally, the average error probability is expressed by

$$P_{e_{d_M}} = P_0 p_{10} + P_1 p_{01} \tag{3.50}$$

where P_0 and P_1 are the probabilities of occurrence of bits 0 and 1, respectively.

3.4.2.2 Detection by the Sample Variance

In the previous section the average luminance of each element c_i was used as the detection metric. However, by exploiting physical characteristics of the PS channel, other statistics of the received signal can also be employed in the detection process.

Section 3.1.2 briefly describes the ordered dithering halftoning algorithm. When printing regular binary text (i.e. black text on white background), halftoning is not employed, and all characters are printed in full black. However, when a grey level luminance is added to the text, the printed characters are composed by black and white dots, increasing the variance of a character. The noise-free output variance as a function of the input luminance of a halftoned character is illustrated in Fig. 3.7a. In this section this characteristic of the halftoning is used, by employing the variance of character c_i as a detection metric.

The NP detector for identifying an abrupt change in the variance of Gaussian noise is simply the sample variance of the observed region. By mapping the (m, n) coordinates to an one-dimensional notation, consider the sample variance as a new detection metric d_{Vi} for element i, given by:

$$d_{Vi} = \frac{1}{N_i} \sum_{n=1}^{N_i} [y_i(n) - \bar{y}_i]^2 \tag{3.51}$$

where N_i is the number of pixels in element i and $y_i(n)$ is the printed and scanned version of $s_i(n)$, according to the PS model described in (3.41). To find the bit error probability when using the variance as the detection metric, it is necessary to estimate the shape and parameters of the probability density function of the random variable d_V. As the sum of N random variables v_i normally distributed according to $z = \sum_{i=1}^{N} v_i^2$ obeys a chi-squared distribution with N degrees of freedom. Similarly, the detection metric d_V in (3.51) also follows a chi-squared distribution assuming that y_i is approximately normally distributed. In this application, however, N corresponds to the number of pixels in the scanned character and it is usually large ($N \gg 100$), and d_V can be assumed as normally distributed, since the chi-squared distribution becomes Gaussian as N becomes large. In the following the expected value and the variance of d_V are presented.

The expected value and the variance of the detection metric d_V in (3.51) can be shown as

$$\mu_{d_V} = [(\mu_\alpha^2 + \sigma_\alpha^2)\sigma_{\eta_2}^2 + \sigma_{\eta_1}^2] r_h(0) + \sigma_{\eta_3}^2 \tag{3.52}$$

and

$$\sigma_{d_y}^2 = \sigma_{\eta_2}^4 r_h^2(0)\left(2\sigma_\alpha^4 + 4\sigma_\alpha^2 \mu_\alpha^2 + \frac{2A}{N}\right) + \frac{2\sigma_{\eta_1}^4 r_h^2(0)}{N} + \frac{2\sigma_{\eta_3}^4}{N}$$
$$+ \frac{4}{N}\left(B\sigma_{\eta_2}^2 \sigma_{\eta_1}^2 r_h^2(0) + B\sigma_{\eta_2}^2 \sigma_{\eta_3}^2 r_h(0) + \sigma_{\eta_1}^2 \sigma_{\eta_3}^2 r_h(0)\right) \tag{3.53}$$

respectively, where $A = 3\sigma_\alpha^4 + 6\sigma_\alpha^2 \mu_\alpha^2 + \mu_\alpha^4$ and $B = \sigma_\alpha^2 + \mu_\alpha^2$. In practice, when scanning is performed at high resolution, N is very large and α and the halftone noise η_2 are the predominant noise sources. In this case, (3.53) can be approximated to

$$\sigma_{d_y}^2 = \sigma_{\eta_2}^4 r_h^2(0)\left(2\sigma_\alpha^4 + 4\sigma_\alpha^2 \mu_\alpha^2\right) \tag{3.54}$$

Using d_V, the conditional error probability p_{01} is given by $p_{01} = \frac{1}{2}\text{erfc}\left(\frac{\mu_{d_V/1} - \lambda_V}{\sqrt{2\sigma_{d_V/1}^2}}\right)$ where $\mu_{d_V/1}$ and $\sigma_{d_V/1}^2$ are the expected value and the variance of d_V when bit 1 is transmitted, respectively, illustrated in Fig. 3.21. Similarly, $p_{10} = \frac{1}{2}\text{erfc}\left(\frac{\lambda_V - \mu_{d_V/0}}{\sqrt{2\sigma_{d_V/0}^2}}\right)$. The average error probability is expressed by

$$P_{e_{d_V}} = P_0 p_{10} + P_1 p_{01} \tag{3.55}$$

3.4.3 Detection Using Higher Order Statistical Moments

Similarly to the variance, higher order statistical moments can also be used to detect different symbols. An approximation of the effect of the assumed PS channel over the these moments is described next.

3.4.3.1 Sample Skewness

The expected value of the sample skewness of a scanned symbol y is given by

$$\mu_{\gamma_{1y}} = E\left\{ \frac{1}{\sigma_y^3 N} \sum_{n=1}^{N} [\alpha w + \alpha \eta_2(n) \circledast h(n) + \eta_1(n) \circledast h(n) + \eta_3(n) - \bar{y}]^3 \right\}$$

$$= \frac{1}{\sigma_y^3 N} \sum_{n=1}^{N} E\{[\alpha \eta_2(n) \circledast h(n) + \eta_1(n) \circledast h(n) + \eta_3(n)]^3\}$$

(3.56)

Recalling that η_1, η_2 and η_3 are zero-mean mutually independent random variables and that third order moments of independent and identically distributed zero-mean random variables are zero, (3.56) becomes

$$\mu_{\gamma_{1y}} = \frac{1}{\sigma_y^3} E\{\alpha\} E\{[\eta_2(n) \circledast h(n)]\}$$

(3.57)

The term $E\{[\eta_2(n) \circledast h(n)]\}$ in equation above, can be shown as

$$\mu_{\gamma_{1y}} = \frac{1}{(\sigma_y^2)^{3/2}} (3\sigma_\alpha^2 \mu_\alpha + \mu_\alpha^3)[(1-w)(-w)^3 + (1-w)^3 w]h_3$$

(3.58)

where σ_y^2 is described by (3.52) and h_3 is given by:

$$h_3 = \sum_{k=-\infty}^{\infty} \sum_{l=-\infty}^{\infty} \sum_{r=-\infty}^{\infty} h(k)h(l)h(r)$$

(3.59)

3.4.3.2 Sample Kurtosis

The sample kurtosis of a scanned symbol is given by

$$\mu_{\gamma_{2y}} = E\left\{ \frac{1}{\sigma_y^4 N} \sum_{n=1}^{N} [\alpha w + \alpha \eta_2(n) \circledast h(n) + \eta_1(n) \circledast h(n) + \eta_3(n) - \bar{y}]^4 \right\}$$

$$= \frac{1}{\sigma_y^4 N} \sum_{n=1}^{N} E\{[\alpha \eta_2(n) \circledast h(n) + \eta_1(n) \circledast h(n) + \eta_3(n)]^4\}$$

$$= \frac{1}{\sigma_y^4 N} \sum_{n=1}^{N} E\{\alpha^4 [\eta_2(n) \circledast h(n)]^4\} + 6E\{\alpha^2 [\eta_2(n) \circledast h(n)]^2 [\eta_1(n) \circledast h(n)]^2\}$$

$$+2E\{\alpha^2 [\eta_2(n) \circledast h(n)]^2 \eta_3^2(n)\} + E\{[\eta_1(n) \circledast h(n)]^4\}$$

$$+6\{[\eta_1(n) \circledast h(n)]^2 \eta_3^2(n)\} + E\{\eta_3^4(n)\} \tag{3.60}$$

The term $E\{[\eta_2(n) \circledast h(n)]^4\}$ in equation above, can be shown as

$$\mu_{\gamma_{2y}} = \frac{1}{\sigma_y^4 N} \Bigg((3\sigma_\alpha^4 + 6\sigma_\alpha^2 \mu_\alpha^2 + \mu_\alpha^4)[(1-w)(-w)^4 + (1-w)^4 w]h_4$$

$$+ 6(\sigma_\alpha^2 + \mu_\alpha^2)\sigma_{\eta_1}^2 \sigma_{\eta_2}^2 r_h^2(0) + 6(\sigma_\alpha^2 + \mu_\alpha^2)\sigma_{\eta_2}^2 \sigma_{\eta_3}^2 r_h(0) + 3\sigma_{\eta_1}^4 r h^2(0)$$

$$+ 6\sigma_{\eta_1}^2 \sigma_{\eta_3}^2 r_h(0) + 3\sigma_{\eta_3}^4 \Bigg)$$

$$\tag{3.61}$$

where h_4 is given by:

$$h_4 = \sum_{k=-\infty}^{\infty} \sum_{l=-\infty}^{\infty} \sum_{r=-\infty}^{\infty} \sum_{s=-\infty}^{\infty} h(k)h(l)h(r)h(s) \tag{3.62}$$

3.4.4 Spectral Based Detection

As discussed in Sect. 3.1.2, the coefficients in the dithering matrix D_H have a direct effect on the quality of the halftone image. Two common eye-pleasing dither matrices structures are those based on green- and blue-noise halftoning models.

The color names for different types of noise are derived from a loose analogy between the spectrum of frequencies present in the noise and the equivalent spectrum of light wave frequencies [51]. In this interpretation, if the pattern of "blue noise," for example, were translated into light waves, the resulting light would be blue.

Therefore, green noise models produce an output formed mostly of midfrequencies spectral components. Blue noise models produce an output formed mostly of high-frequency components. These two kinds of halftoning models are applied in this work for text watermarking.

Using TLM, text characters have their luminances modified to convey information. The printed version of the modified characters can be halftoned with different matrices D_H. Characters of equivalent luminances printed using different D_H's present approximately the same average luminance after printing, however the spectral characteristics are significantly different. An example of employing different halftone matrices is given in Fig. 3.22. This scheme is referred to as text halftone modulation (THM).

Fig. 3.22 Exaggerate illustration of different halftone patterns

Using fundamentals of classical spectral estimation theory, a possible detection metric to classify a given character is to use sub-band spectral features. For this task, the power spectral density (PSD) of a scanned character $y(n)$ of size N is divided into L subbands of size W, where $W = N/L$. The average power of each of these subbands represents one among L features.

Let the average power of l-th sub-band be given by

$$d_l = \frac{1}{W} \sum_{\omega=W(l-1)}^{lW-1} |Y(\omega)|^2 \tag{3.63}$$

where $Y(\omega)$ is

$$Y(\omega) = \frac{1}{\sqrt{N}} \sum_{n=0}^{N-1} y(n)e^{-j2\pi n\omega/N}$$

$$= \frac{1}{\sqrt{N}} \sum_{n=0}^{N-1} \left\{ \{\alpha[w + \eta_2(n)] + \eta_1(n)\} \circledast h_{ps}(n) + \eta_3(n) \right\} e^{-j2\pi n\omega/N} \tag{3.64}$$

To determine to which class the received vector y belongs, the average power of each sub-band l is used as a feature. Therefore, the feature classification vector is given by

$$\mathbf{d} = [d_1 \quad d_2 \quad \dots \quad d_l]^\top \tag{3.65}$$

The expected value μ_{d_l} of a feature d_l is given by:

$$\mu_{d_l} = E\{d_l\}$$

$$= E\left\{ \frac{1}{W} \sum_{\omega=W(l-1)}^{lW-1} |Y(\omega)|^2 \right\}$$

$$= E\left\{ \frac{1}{W} \sum_{\omega=W(l-1)}^{lW-1} \frac{1}{(N^2)} \sum_{n=0}^{N-1} \{\alpha w \circledast h_{ps}(n) + \alpha \eta_2(n) \circledast h_{ps}(n) \right. \tag{3.66}$$

$$+ \eta_1(n) \circledast h_{ps}(n) + \eta_3(n)\} e^{-j2\pi n\omega/N} \sum_{m=0}^{N-1} \{\alpha w \circledast h_{ps}(n)$$

$$\left. + \alpha \eta_2(n) \circledast h_{ps}(n) + \eta_1(n) \circledast h_{ps}(n) + \eta_3(n)\} e^{j2\pi n\omega/N} \right\}$$

Considering the statistical characteristics (zero-mean, mutually independent) assumed for the noise terms in (3.67), the cross terms are canceled and the expected value in (3.67) becomes

$$\mu_{d_l} = \frac{1}{W} \sum_{\omega=W(l-1)}^{lW-1} \alpha^2 w^2 H_{ps}(\omega) H_{ps}^*(\omega) \delta(\omega)/N$$
$$+ \alpha^2 H_{ps}(\omega) H_{ps}^*(\omega) \sigma_{\eta_2}^2 + H_{ps}(\omega) H_{ps}^*(\omega) \sigma_{\eta_1}^2 + \sigma_{\eta_3}^2 \tag{3.67}$$

where $\delta(\cdot)$ is the unit impulse function. One can show that, when N is large, the variance $\sigma_{d_l}^2 = E\{d_l^2\} - \mu_{d_l}^2$ of a feature d_l can be approximated to:

$$\sigma_{d_l}^2 = \frac{1}{W^2} \sum_{\omega=W(l-1)}^{lW-1} \sum_{v=W(l-1)}^{lW-1} 3\sigma_{\eta_2}^4 (3\sigma_\alpha^4 + 6\sigma_\alpha^2 \mu_\alpha^2 + \mu_\alpha^4)$$

$$|H_{ps}(\omega)|^2 |H_{ps}(v)|^2 + 3\sigma_{\eta_1}^4 (\sigma_\alpha^2 + \mu_\alpha^2) |H_{ps}(\omega)|^2 |H_{ps}(v)|^2 + 3\sigma_{\eta_3}^4 - \mu_{d_l}^2$$

$$\tag{3.68}$$

Assuming that d_l is normally distributed, from μ_{d_l} and $\sigma_{d_l}^2$ the theoretical detection error rates are determined, which are presented in the experiments section of this chapter.

3.4.5   Combining Different Metrics

This section proposes and discusses improvements by using additional detection metrics in the system, and by combining the results of these metrics into a new decision criterion. This approach falls into a multicriteria classification problem, where each element c_i must be classified as belonging to one among S classes by determining an estimated $\hat{\omega}$, $\hat{\omega} \in \Omega$.

Consider, for example, two of the metrics discussed so far: the mean d_M and the variance d_V. There is not a deterministic relationship between the luminance level (DC) and the variance level in a printed and scanned character. Hence, a better strategy than using separately either the mean luminance or the sample variance of element c_i as a detection metric, is to combine these metrics into a new single decision statistic. Several techniques [26] can be used to merge different metrics, and this work adopts the Bayes Classifier [26], a classical method which yields the minimum average error for Gaussian distributed patterns.

In this case, in contrast to Eqs. (3.42) and (3.51), where the decision for estimated $\hat{\omega}$ depends on a single value, decision is based on the vector.

$$\mathbf{d}_i = \begin{bmatrix} d_{Mi} \\ d_{Vi} \end{bmatrix} \tag{3.69}$$

According to the results derived in Sect. 3.4.2.1, the PDF of the detection metric d_M is described by:

$$p(d_M) = \frac{1}{\sqrt{2\pi\sigma_{d_M}^2}} \exp\left[-\frac{1}{2\sigma_{d_M}^2}(d_M - \mu_{d_M})^2 \right] \tag{3.70}$$

Considering the $S = 2$ (or 1 bit) case, let $\mu_{d_M} = \mu_{d_M/0}$ and $\sigma_{d_M}^2 = \sigma_{d_M/0}^2$ be the expected value and the variance of d_M when bit 0 is transmitted. Correspondingly, let $\mu_{d_M} = \mu_{d_M/1}$ and $\sigma_{d_M}^2 = \sigma_{d_M/1}^2$ be the expected value and the variance of d_M when bit 1 is transmitted. Therefore, the conditional PDF's of d_M given that bit 0 and bit 1 were transmitted are described by

$$p(d_M/b = 0) = \frac{1}{\sqrt{2\pi\sigma_{d_M/0}^2}} \exp\left[-\frac{1}{2\sigma_{d_M/0}^2}(d_M - \mu_{d_M/0})^2 \right] \tag{3.71}$$

and

$$p(d_M/b = 1) = \frac{1}{\sqrt{2\pi\sigma_{d_M/1}^2}} \exp\left[-\frac{1}{2\sigma_{d_M/1}^2}(d_M - \mu_{d_M/1})^2 \right] \tag{3.72}$$

respectively. A plot with the conditional probability density functions described by the above equations is given in Fig. 3.20. The larger deviation noticeable in the leftmost curve (bit 1), is due to the signal dependent noise term, η_2.

Equivalently, from the results of Sect. 3.4.2.2, where the detection metric used was the variance (in (3.51)), Fig. 3.21 plots the corresponding probability density functions. The parameters used to generate this plot were identical to those of Fig. 3.20, and similarly to that figure, the larger deviation noticeable in the rightmost curve (bit 1), is due to the signal dependent noise term, η_2.

3.5 A Practical Authentication Protocol

This section discusses an interesting document authentication system [10, 59] that can be applied to electronic and printed documents, possibly to be used in conjunction with the traditional methods mentioned in previous chapters. The system is based on TLM and as such it can be set to cause a very low perceptual impact. Unlike a digital signature, which protects the binary codes of the documents, the system proposed here protects the visual content. In contrast to digital watermarking schemes that transmit a hidden message, the proposed system classifies the document as authentic or non-authentic.

An advantage of the system is that it does not require a database to store information to be compared. For this reason, the proposed system is coined text self authentication (TSA) [10]. Moreover, special hardware is not required, except for a consumer scanner when printed documents are considered. Notice that TSA does not rely on a specific function to modify each character, which can be either performed with very low perceptual impact using text watermarking techniques discussed in this work, or visibly to increase robustness.

Two applications scenarios are considered. In the first scenario, TSA is described in a noise-free environment and the false alarm rate (i.e., the probability of assigning an authentic document as non-authentic) is zero. In the second scenario, it is assumed that errors may occur in the detection of the modified character feature, mainly due to the noise in the print and scan (PS) process. In this case, a correlation-based detector is proposed and an analysis is performed to determine the detection error probability. Applications include passports, driver's licenses and ID cards, and legal notes.

The proposed framework for authentication scrambles the binary representation of the original text string with a key that depends on the string. The resulting scrambled vector is used to create another vector of dimension equal to the number of characters in the document. This is used as a rule to modulate each character individually, according to the TLM approach.

Related approaches for image authentication in which a digital watermark is generated with a key that is a function of some feature f in the original image has been proposed in the literature, as in [18, 33, 62], for example. To avoid that f be modified by the embedding of the watermark itself, hence frustrating the watermark detection process, only characteristics of a portion of the image must be used. It is possible, for example, to extract features from the low-frequency components, and to embed the watermark in the high frequency components, as discussed in [19].

In contrast, in TSA the modified characters luminance do not alter the feature used to generate the permutation key, which are the characters "meanings." The system is described in the following.

3.5.1   Encryption

- Let vector $\mathbf{c} = [c_1, c_2, \ldots, c_K]$ of size K represent a text string with K characters.
- Let vector $\mathbf{s} = [s_1, s_2, \ldots, s_K]$ represent the luminances of characters $[c_1, c_2, \ldots, c_K]$, respectively.
- Let $c_i \in \Phi$ ($\Phi = \{$a,b,c, $\ldots$, X, Y, Z $\}$, for example), where Φ has cardinality $|\Phi|$.
- Let c_{bi} be the binary representation of c_i.
- Let $\mathbf{c}_b$ be the binary representation of $\mathbf{c}$, where $\mathbf{c}_b$ has size $|\mathbf{c}_b| = K \log_2 |\Phi|$.

- Let $\kappa = f(\mathbf{c}_b)$ be a function of $\mathbf{c}_b$. κ is used as a key to generate a pseudo-random sequence (PRS) $\mathbf{k}$, such that the PRS's are ideally orthogonal for different keys κ.
- Let $\mathbf{c}'_b = \mathbf{c}_b \oplus \mathbf{k}$, where $\oplus$ represents the "exclusive or" (XOR) logical operation.
- Let $\mathcal{M}$ be a function that maps $\mathbf{c}'_b$, with $|\mathbf{c}_b|$ bits, to another vector $\mathbf{w}$, with K bits.

In order to provide security, $\mathbf{w}$ is encrypted with the private key of a public-key cryptosystem [28, 49, 51]. Public-key cryptosystems use two different keys, one for encryption, κ_e, and one for decryption, κ_d. The private key κ_e is only available for users who are allowed to perform the authentication process. On the other hand, anyone can have access to the public key κ_d to *only check* whether a document is authentic, without the ability to generate a new authenticated document.

Let $\mathbf{w}_e$ be the encrypted version of $\mathbf{w}$ based on the key κ_e, using a public key encryption scheme such as the RSA [51], for example. To authenticate the text document, vector $\mathbf{s}$ (which represents the luminance of the characters in the document) is modified such that $\mathbf{s} = \mathbf{w}_e$. Therefore, the document is authenticated by setting the luminance of each character c_i equal to s_i (Fig. 3.23).

Fig. 3.23 Encryption block diagram. Block 's/b' represents string-to-binary conversion. Block '$\mathcal{M}$' represents a mapping of $\mathbf{c}'_b$ from $|\mathbf{c}_b|$ bits to K bits. The symbol $\oplus$ represents the "exclusive or" (XOR) logical operation, ©2009 IEEE. Reprinted from [12] with permission license no. 4221170750957

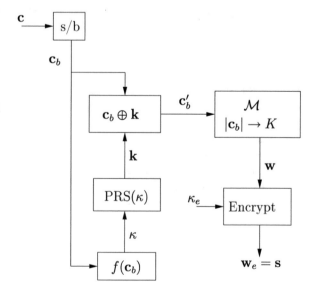

Fig. 3.24 Extracting $\hat{\mathbf{c}}$ and $\hat{\mathbf{w}}$ from the received document

3.5.2   Decryption

In the verification process, OCR is applied to the document in the printed cases. In addition, the average luminance of each character is determined, as illustrated in Fig. 3.24. Therefore, when testing for the authenticity of the document one has access to a received $\hat{\mathbf{c}}$ and a received $\hat{\mathbf{s}}$, where $\hat{\mathbf{c}}$ and $\hat{\mathbf{s}}$ represent the received vectors $\mathbf{c}$ and $\mathbf{s}$, respectively. It is assumed that the conditions are controlled such that no OCR or luminance detection errors occur. Moreover, one has access to a public key κ_e for decryption in the RSA algorithm and a permutation key $\kappa = f(\hat{\mathbf{c}}_b)$, which depends on $\hat{\mathbf{c}}$.

Using the public key κ_d, it is possible to decrypt $\hat{\mathbf{s}} = \hat{\mathbf{w}}_e$ into $\hat{\mathbf{w}}$. Using κ, it is possible to scramble $\hat{\mathbf{c}}_b$ (the binary representation of $\hat{\mathbf{c}}$) yielding $\hat{\mathbf{c}}'_b$. Applying the same mapping rule $\mathcal{M}$ of the encryption process to $\hat{\mathbf{c}}'_b$ yields a new vector $\hat{\mathbf{w}}'$.

If $\hat{\mathbf{w}}' = \hat{\mathbf{w}}$ the document is assumed authentic. Else, it is assumed that one or more characters have been altered. A block diagram of the authentication test process is given in Fig. 3.25.

If an attacker changes one or more characters in the document such that $\hat{\mathbf{c}} \neq \mathbf{c}$, $\hat{\mathbf{w}}$ and $\hat{\mathbf{w}}'$ are two completely different sequences (quasi-orthogonal) with very high probability, failing the authentication test.

Although in the above description OCR has been included in the detection process assuming that the document has been printed and scanned, the proposed authentication protocol can be applied to digital documents.

3.5.3   Noisy Environment

The final stage of TSA illustrated in Fig. 3.25 performs an equality test to decide whether a document is authentic. In noisy environments such as the PS channel, however, this approach can cause a high false alarm rate, that is, documents can be wrongly claimed as non-authentic.

Experimental analyses [9, 60] indicate that the result of the evaluation of the average luminance d_M in TLM is approximately normally distributed, such that $\hat{\mathbf{w}} = \mathbf{w} + \mathbf{n}$, where $\mathbf{n}$ is additive white Gaussian noise (AWGN). A normal distribution can also be assumed for result of the detection in line shift coding algorithms [39].

Therefore, instead of using an equality test, a correlation detector to check the similarity between the expected $\hat{\mathbf{w}}$ and the received $\hat{\mathbf{w}}'$ is used. In this case, the

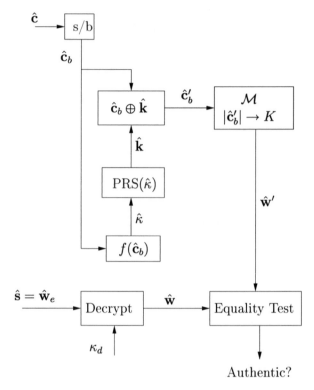

Fig. 3.25 Decryption block diagram. Block 's/b' represents string-to-binary conversion. Block '$\mathcal{M}$' represents a mapping of $\hat{\mathbf{c}}'_b$ from $|\mathbf{c}_b|$ bits to K bits. The symbol $\oplus$ represents the "exclusive or" (XOR) logical operation, ©2009 IEEE. Reprinted from [12] with permission license no. 4221170750957

document is assumed authentic if the result of the linear correlation T_a between $\hat{\mathbf{w}}$ and $\hat{\mathbf{w}}'$ is greater than a given threshold λ_a. Linear correlation is employed as it is the optimal robustness metric when the channel noise can be modeled as AWGN [5].

Therefore, in the proposed correlation test the document is assumed authentic if

$$T_a = \text{Cor}(\hat{\mathbf{w}}, \hat{\mathbf{w}}') = \frac{1}{K} \sum_{i=1}^{K} \hat{w}_i \hat{w}'_i > \lambda_a \tag{3.73}$$

where T_a is a normally distributed random variable and $\hat{w}_i$ and $\hat{w}'_i$, $i = 1, \ldots, K$ are the elements in $\hat{\mathbf{w}}$ and $\hat{\mathbf{w}}'$, respectively. When the document is authentic, vector $\hat{\mathbf{w}}'$ is given by $\hat{\mathbf{w}}' = \hat{\mathbf{w}} + \mathbf{n}$. When the document is tampered with, it is expected that

$$\hat{\mathbf{w}} \perp \hat{\mathbf{w}}' \therefore E\left\{\frac{1}{K} \sum_{i=1}^{K} \hat{w}_i \hat{w}'_i\right\} = 0 \tag{3.74}$$

where the operator $\perp$ represents statistical orthogonality. Assuming that $\mathbf{n}$ and $\hat{\mathbf{w}}$ are distributed according to $\mathbf{n} \sim \mathcal{N}(0, \sigma_n^2)$ and $\hat{\mathbf{w}} \sim \mathcal{N}(\mu_w, \sigma_w^2)$, the expected value in (3.73) is given by

$$\mu_{T_a} = E\left\{\frac{1}{K}\sum_{i=1}^{K}\hat{w}_i(\hat{w}_i + n_i)\right\}$$

$$= \frac{1}{K}\sum_{i=1}^{K}E\{\hat{w}_i^2\} + E\{\hat{w}_i n_i\} \tag{3.75}$$

$$= \mu_w^2 + \sigma_w^2$$

The variance of (3.73) is given by

$$\sigma_{T_a}^2 = E\left\{\left(\frac{1}{K}\sum_{i=1}^{K}\hat{w}_i(\hat{w}_i + n_i)\right)^2\right\} - \mu_T^2$$

$$= \frac{1}{K^2}\sum_{i=1}^{K}\sum_{j=1}^{K}E\{(\hat{w}_i^2 + \hat{w}_i n_i)(\hat{w}_j^2 + \hat{w}_j n_j)\} - \mu_T^2$$

$$= \frac{1}{K^2}\sum_{i=1}^{K}\sum_{j=1}^{K}E\{\hat{w}_i^2\hat{w}_j^2 + \hat{w}_i^2\hat{w}_j n_j + \hat{w}_i n_i\hat{w}_j^2 + \hat{w}_i n_i\hat{w}_j n_j\} - \mu_T^2 \tag{3.76}$$

$$= \mu_w^4 + 2\mu_w^2\sigma_w^2 + \sigma_w^4 + \frac{1}{K}(\mu_w^2 + \sigma_w^2)\sigma_n^2 - \mu_{T_a}^2$$

$$= \frac{(\mu_w^2 + \sigma_w^2)\sigma_n^2}{K}$$

The conditional error probability p_0 given that the document is tampered with is $p_0 = \Pr(T_a > \lambda_a|\text{tampered})$, where λ_a is a decision threshold. Using the complementary error function erfc, $p_0 = \frac{1}{2}\text{erfc}\left(\frac{\lambda_a - \mu_{T_a/0}}{\sqrt{2\sigma_{T_a/0}^2}}\right)$, where $\mu_{T_a/0}$ and $\sigma_{T_a/0}^2$ are respectively the mean and the variance of T_a for tampered with documents. Equivalently, if the document is authentic, the conditional error probability is given by $p_1 = \frac{1}{2}\text{erfc}\left(\frac{\mu_{T_a/1} - \lambda}{\sqrt{2\sigma_{T_a/1}^2}}\right)$, where $\mu_{T_a/1}$ and $\sigma_{T_a/1}^2$ are respectively the mean and the variance of T_a for authentic documents. Finally, the average error probability is expressed by

$$P_{e_{TSA}} = P_0 p_0 + P_1 p_1 \tag{3.77}$$

where P_1 and P_0 are the probabilities of occurrence of authentic and tampered with documents, respectively.

3.6   Position Based Coding

This section introduces an alternative coding scheme for TLM, which causes a reduced distortion to the text while maintaining the transmitting rate. The scheme is also applicable to other methods based on the modification of individual characters, discussed earlier in this chapter.

3.6.1   Sequential Modulation

A characteristic that is common to TLM and all the hardcopy watermarking techniques described in this chapter is that the modification of the characters (or other structures) are performed in sequence, according to the bit to be embedded. Notice that in TLM each bit is encoded into the characters according to the order of appearance of each character in the document. For this reason, this scheme is referred to in this section as sequential modulation (SM), as opposed to position based coding (PBC), described in the following.

3.6.2   Positional Encoding

Using PBC, information is related to the position of the modulated characters in the document. A related method focused on digital images has been proposed in [3] and generalized in [8], where the authors embed information in an image by adding to it pseudo-random blocks in different positions, according to the information.

Using the positional encoding, the information to be embedded is related to the *position* of a given number of modulated characters in the document. Consider a document with K characters. A message **b** is embedded into the document by modulating Q different characters in the text, where $Q < K$. Therefore, the set of indexes i that contain the modulated characters represent the embedded information. For instance, using an appropriate coding rule, a bit string $\mathbf{b} = [11010]$ can be embedded into the text:

POSITIONAL ENCODING

by modulating 2 characters ($Q = 2$) such that the watermarked text becomes

POSITIONAL ENCODING

A different bit string $\mathbf{b} = [01110]$ could cause

POSITIONAL ENCODING

for example. An efficient coding rule mapping "input information" ↔ "output positions" is described in [8]. Notice that this coding rule does not use a preset codebook, which would be computationally expensive. Instead, a mathematical relationship between input and output is used.

3.6.3   Distortion of PBC Versus SM

In the analysis of this section, assume that $S = 2$, representing the number of modulation levels. Let R_P be the embedding capacity in a document using PBC, representing the number of bits embeddable. R_P depends on K and on Q. It is given by:

$$R_P = \log_2 \binom{K}{Q} \tag{3.78}$$

where

$$\binom{K}{Q} = \frac{K!}{Q!\,(K-Q)!} \tag{3.79}$$

is the binomial coefficient in combinatorial analysis [48]. The payload given in (3.78) is illustrated as a function of K and Q in Fig. 3.26.

In contrast, the embedding capacity using SM is simply a function of K, given by

$$R_S = K \tag{3.80}$$

In order to compare the embedding capacities of PBC and SM, (3.78) and (3.80) must be expressed as a function of the same parameters for the same distortion level.

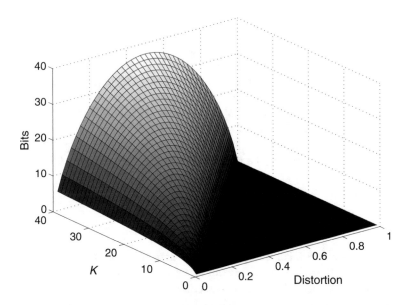

Fig. 3.26 Capacity of PBC as function of K and of the distortion D, ©2009 IEEE. Reprinted from [11] with permission license no. 4221160975332

Let Q be the number of modified characters. Let the amount of distortion in a modulated character be represented by ϑ. In this section, let the average distortion caused by embedding a message in a document be represented by ε, given by

$$\varepsilon = \frac{Q\vartheta}{K} \tag{3.81}$$

For simplicity, consider $\vartheta = 1$ for the rest of this section.

Although the message **b** to be embedded is defined by the user, consider **b** as a realization of a random process. In PBC, Q is deterministic, independent of the bit string **b**. In contrast, using SM, the distortion depends on the embedded message. For instance, the distortion caused by **b** = [111101] is stronger than that caused by **b** = [000010].

Assume that bit 0 and bit 1 in **b** occur with equal probability, such that

$$p_0 = p_1 = 0.5 \tag{3.82}$$

where p_0 and p_1 are the probabilities of occurrence of bits 0 and 1, respectively. Using SM, p_0 and p_1 translate directly into the probabilities of occurrence of non-modulated and modulated characters. In a document composed of K characters, the expected number of modulated characters (or 'bit 1' characters) using SM is

$$E\{Q\} = Kp_1 = 0.5K \therefore K = 2E\{Q\} \tag{3.83}$$

Because in PBC Q is deterministic, $E\{Q\} = Q$. Using this result, the substitution of (3.83) into (3.80) yields

$$R_S = 2Q \tag{3.84}$$

Figure 3.27 shows the capacities of both methods, illustrating the overall better performance of PBC, for $K = 200$. This figure shows that if 20 out of the 200 characters in the document are modulated ($\varepsilon = 20/200 = 0.1$), for example, PBC encodes 90 bits whereas SM encodes 40 bits.

A surface corresponding to the ratio R_P/R_S is given in Fig. 3.28. A two dimensional representation of the ratio is given in Fig. 3.29, for $K = 200$.

3.6.4 Detection

In the digital only domain, in applications where the document does not suffer significant distortions, the detection error rate is practically zero.

However, as discussed in previous sections, in hardcopy applications the PS channel can be seen as a noisy communications channel, causing several distortions to the image, such as low-pass filtering, non-linear gains, and additive noise.

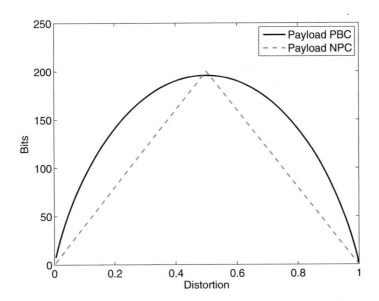

Fig. 3.27 Capacity as a function of the distortion D, for $K = 200$, ©2009 IEEE. Reprinted from [11] with permission license no. 4221160975332

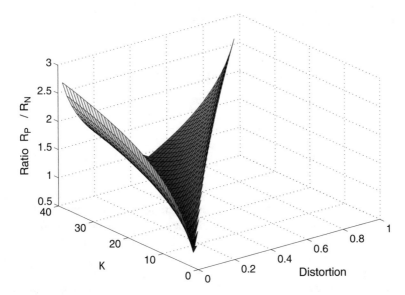

Fig. 3.28 Ratio R_P/R_S, as a function of the distortion D, ©2009 IEEE. Reprinted from [11] with permission license no. 4221160975332

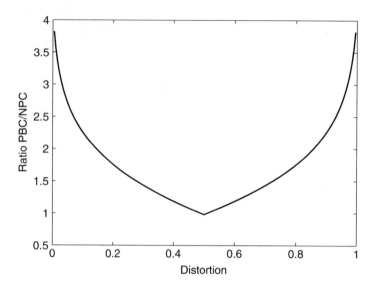

Fig. 3.29 Ratio R_P/R_S, as a function of the distortion D, for $K = 100$, ©2009 IEEE. Reprinted from [11] with permission license no. 4221160975332

The simplest detection metric to determine embedded luminance is the average luminance of the character, given in (3.85). For this reason, this metric is used in the examples of this section.

By mapping the (m, n) coordinates in (3.39) to a one-dimensional notation, the detection metric d_{Mi} for the i-th character is given by:

$$d_{Mi} = \frac{1}{N_i} \sum_{n=1}^{N_i} y_i(n), \tag{3.85}$$

where N_i is the number of pixels in character i and $y_i(n)$ is the printed and scanned version of $s_i(n)$.

In the $S = 2$ case, for example, if the average luminance is above a threshold λ_M (possibly determined through statistical training, for a given PS channel), y_i is assumed as modulated (bit 1). Else, it is assumed as non-modulated (bit 0).

In a hardcopy application, considering the sum of the several distortions of the PS channel, it is observed experimentally that the detector output can be modeled as normal random variable, as supported by the central limit theorem.

With these assumptions, let $d_{M/0}$ be a random variable representing the result of the detection statistic for **non-modulated** characters and assume it is distributed according to $d_{M/0} \sim \mathcal{N}(\mu_{d_{M/0}}, \sigma^2_{d_{M/0}})$. Similarly, let $d_{M/1}$ represent the detection statistic of **modulated** characters, given by $d_{M/1} \sim \mathcal{N}(\mu_{d_{M/1}}, \sigma^2_{d_{M/1}})$.

Because of the noisy channel, a disadvantage of PBC is that if the detection process wrongly assumes a character y_i as modified, the embedded message is entirely lost. In contrast, SM has the advantage that bits can be recovered through error correcting codes, which costs some amount of usable embedding rate.

To reduce this problem in PBC, an alternative detection scheme that does not rely on the threshold λ_M is employed. Instead of evaluating if the average luminance of y_i is greater or smaller than λ_M, the Q characters with the highest average luminance are determined. This causes a significant reduction in the error rate, as shown in the following.

In PBC, only Q among K characters are modulated. The detector selects as the modulated characters only the Q locations which provide the Q highest detection values. Therefore, to determine the error probability, the probability of erroneously assuming the presence of a modulated character in a given location must be determined.

The probability p_e that a **non-modulated** character presents a higher detection value than a modulated character must be determined. It can be shown [7] that this probability is:

$$p_e = \Pr(d_{M/0} > d_{M/1}) \tag{3.86}$$

$$= \frac{1}{2}\mathrm{erfc}\left(\frac{(\mu_{d_{M/0}} - \mu_{d_{M/1}})\sigma_{d_{M/1}}}{(\sigma^2_{d_{M/0}} + \sigma^2_{d_{M/1}})\sqrt{2}\cos\left(\arctan\frac{\sigma_{d_{M/0}}}{\sigma_{d_{M/1}}}\right)}\right) \tag{3.87}$$

For the case where $\sigma_{d_{M/0}} = \sigma_{d_{M/1}}$, (3.87) is reduced to

$$p_e = \frac{1}{2}\mathrm{erfc}\left(\frac{(\mu_{d_{M/0}} - \mu_{d_{M/1}})\sigma_{d_{M/1}}}{2\sigma_{d_{M/0}}}\right) \tag{3.88}$$

Equation (3.87) describes the probability of erroneously detecting a *single* character as modulated. The total error probability must take into account K detections. Notice that erroneously detecting a single character results in missing the entire message. Thus, considering all the characters in the document, the probability of finding the wrong message using PBC is:

$$p_{PBC} = 1 - (1 - p_e)^{Q(K-Q)} \tag{3.89}$$

When the channel noise is strong, using a perceptually transparent modulation causes a very high error rate. To suppress that, the modulation gain can be increased to a visible level. However, using SM, empirical tests indicate that this causes a disturbing pattern on the text. On the other hand, using PBC, only Q characters are modified. Although the modulation is visible, it is localized and does not have affect the readability of the text [13].

3.7   From Black and White to Color

This section extends the TLM concept by using color as a modifiable feature. The idea of text color modulation (TCM) as a text authentication method has been thoroughly discussed in literature [13, 22, 58]. This section presents a detection metric and an analysis determining the detection error rate in TCM, considering the assumed PS channel model. In addition, a perceptual impact model is employed to evaluate the perceptual difference between a modified and a non-modified character. Combining this perceptual model and the results from the detection error analysis it is possible to determine the optimal color modulation values. Based on the TCM approach, this chapter provides the following contributions:

1. Motivation and the benefits of TCM over other text modulation methods are explained.
2. Considering the PS channel model, a detection metric for TCM is proposed, which combines information of different color channels. For this metric, an analysis to estimate the theoretical error rate is presented. The Bhattacharyya bound [23] is used to determine the error probability when the information from the blue and red channels are combined.
3. Using the perceptual CIE L*a*b* color space [30], a color perceptual difference function is used to determine the perceptual impact of the modulation in each color channel, considering the modulation in only one channel or combining the channels.
4. Using the results from the error analysis and from the perceptual impact, an optimum modulation level is achieved, considering a tradeoff between error rate and perceptual impact.
5. In color printing, each color channel is printed in a given orientation angle to minimize the interference between color channels [4]. This is exploited in this work, as the detection of the dominant angle in each character is also used to help identify the type of color modulation employed, reducing the detection error rate.

3.7.1   Motivation for Using Color

An obvious additional requirement of TCM in comparison to TLM and THM, is that a color printing device is necessary. However, several applications require the use of color printing, such as banking documents, legal notes, identification cards, among others, opening a wide range of applications for TCM.

Several advantages regarding the performance of the system also motivate the use of TCM. In comparison to TLM, for example, TCM presents a lower detection error rate, given a perceptual impact. This occurs because the information of different color channels can be combined, reducing the error rate. Moreover, although both the luminance and chrominance frequency responses of the human vision system (HVS) have low pass characteristics, the spatial sensitivity of the human eye to variations in chrominance fall off faster as a function of increasing spatial frequency than does the response to spatial variations in luminance [4]. This means that the chrominance channel has a narrower bandwidth than the luminance channel. This is relevant because printing uses halftoning [4], and therefore more chromatic error than luminance error is allowed. Aiming at designing a low perceptual error halftoning algorithm, the possibility of increasing the perceived smoothness of halftone textures by trading errors in the luminance channel for errors in the chrominance channel was first pointed out in [44].

Notice that extensions of TLM that exploit characteristics of halftoning to improve detection have been proposed, such as THM and higher order statistics detection. Because color printing also uses halftoning, these methods can also be extended to the color case, providing additional reduction in the error rate. Also, it is important to note that other document authentication methods such as line spacing modulation and character position [16, 29] can be used in parallel with TCM.

3.7.2 Some Comments on Color Printing

The most common digital color standard combines three color channels—red, green, and blue (RGB)—to represent colors [26]. However, because printing is usually performed on a white surface, it is necessary to invert each of these separations. When a negative image of the red component is produced, the resulting image represents the cyan component of the image. Likewise, negatives of the green and blue components produce magenta and yellow separations, respectively. This is done because cyan, magenta, and yellow (CMY) are subtractive primaries which each represent two of the three additive primaries (RGB) after one additive primary has been subtracted from white light.

When CMY are combined in the printing process, the result should be a perceptually acceptable reproduction of the original image. However, in practice, due to limitations in the ink pigments, the darker colors usually present a muddied appearance. To reduce this effect, a black separation is also created, which improves the shadow and contrast of the image. Numerous techniques exist to derive this black separation from the original image; these include grey component replacement, under color removal, and under color addition. This printing technique is referred to as cyan, magenta, yellow and black (CMYK), where K is short for key, which in this case is black.

Although CMYK is used in the printing process, after the image is scanned, a conversion back to RGB is possible, ideally resembling the original RGB representation prior to printing.

In the case of color printing, the halftoning algorithm is applied to each color channel separately, and the result is printed over the paper. Undesirable patterns such as Moiré patterns [25] may arise due to interference between two or more of the dithering matrices that represent each of the colorants. In order to reduce this effect, the dithering matrices for each channel are rotated to be placed as far apart

Fig. 3.30 Example of color
halftoning test targets,
illustrating the C, M, Y and K
colorants, ©2009 IEEE.
Reprinted from [13] with
permission license no.
4220870332019

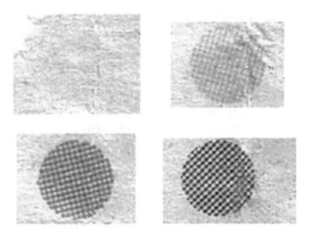

in angle as possible. A typical solution is to place the Y screen at 0°, M at 15°, K at
45° and C at 75° [4]. Figure 3.30 shows a halftone test target, extracted from 'The
Londoner' newspaper. This figure illustrates the printing angle used in halftoning
screen for the C, M, Y and K colorants. This characteristic of color printing is
exploited in this section as an auxiliary metric to help improving the detection in
TCM, as discussed in Sect. 3.7.5.

3.7.2.1 Color PS Channel Analytical Model

In this section a color PS channel model is described that include distortions
occurred in each color channel.

For a given color channel C, $C \in \{R, G, B\}$ (where R, G and B stand for red,
green and blue channels, respectively) the digital scanned image y is represented by

$$y_C(m, n) = g_{sc} \left\{ \{g_{prc}[b_C(m, n)] + \eta_{1c}(m, n)\} \circledast h_{ps}(m, n) \right\} + \eta_{3c}(m, n),$$

(3.90)

where each term corresponds to their equivalent described in (3.7).

Notice that the model in (3.90) describes the input-output relationship of the
color PS channel. Although the printing process uses the C, M, and Y colorants
added to the K colorant to provide proper contrast and color darkness, the input and
the output signals of the process can be described in terms of their RGB components,
as presented in (3.90).

3.7.3 The Detection Process

In this section a detection metric is proposed and an analysis for the error rate is
provided.

For simplicity, the (m, n) coordinate system is mapped to a one dimensional notation, and the index i is dropped from c_i and its derivations. Assuming that the same statistical and distortions assumptions are valid for the color case, the PS channel model becomes

$$y_C(n) = \{\alpha_C[b_{0_C} + \eta_{2_C}(n)] + \eta_{1_C}(n)\} \circledast h_{ps}(n) + \eta_{3_C}(n), \qquad (3.91)$$

which is the model assumed for the error rate analysis.

3.7.3.1 Detection Metric

Human vision experiments illustrate that variations in the green channel are more easily perceived than variations in the red and blue channels [37]. This is supported by experiments performed in the experiments section. For this reason, a modulation scheme which modifies only the blue and red channels is used in this work. The green channel is used as a reference. This increases the robustness of the method, as illumination variations and scanner and printing gains affect all the channels. Although the channels are not affected identically, using a difference metric is more robust than using a simple threshold for the average value of a single channel, as illustrated in the experiments section. Figures 3.31 and 3.32 show the histograms of characters modulated only in the blue and red channels, respectively. Therefore, a non-modulated character is totally black, and a modulated character has its red and blue components modified.

Based on the characteristics of the observed printed and scanned characters, the proposed detection metric is given by

$$\mathbf{d}_2 = [d_b \quad d_r]^\top \qquad (3.92)$$

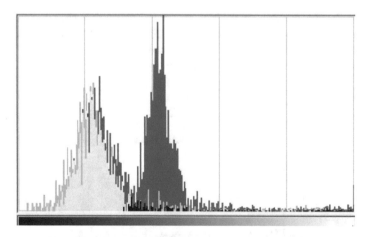

Fig. 3.31 Example of three channel histogram for a character modulated only in the blue channel, ©2009 IEEE. Reprinted from [13] with permission license no. 4220870332019

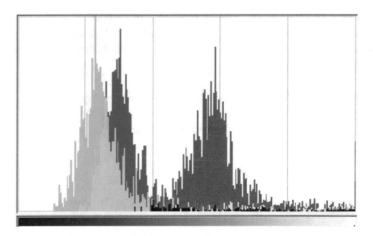

Fig. 3.32 Example of three channel histogram for a character modulated only in the red channel, ©2009 IEEE. Reprinted from [13] with permission license no. 4220870332019

where $\top$ represents the transpose operation and

$$d_b = d_B - d_G \tag{3.93}$$

where

$$d_B = \frac{1}{N} \sum_{n=1}^{N} y_B(n)$$
$$= \frac{1}{N} \sum_{n=1}^{N} \{\alpha_B[w_B + \eta_{2B}(n)] + \eta_{1B}(n)\} \circledast h_{ps}(n) + \eta_{3B}(n), \tag{3.94}$$

represents the average value in the blue channel after printing and w_B represents the modulation level in the blue channel prior to printing, such that $\bar{b}_B = w_B$. Similarly, d_G represents the average value in the green channel.

The expected value of the detection metric d_b is given by

$$\mu_{d_b} = E\{d_b\}$$
$$= E\{d_B - d_G\} \tag{3.95}$$
$$= \mu_{\alpha_B} w_B - \mu_{\alpha_G} w_G$$

The variance of d_b is given by:

$$\sigma_{d_b}^2 = E\{(d_b - \mu_{d_b})^2\}$$
$$= E\left\{\left(\frac{1}{N} \sum_{n=1}^{N} \{\alpha_B[w_B + \eta_{2B}(n)] + \eta_{1B}(n)\} \circledast h_{ps}(n)\right.\right. \tag{3.96}$$

$$+\eta_{3B}(n) - \frac{1}{N}\sum_{n=1}^{N}\{\alpha_G[w_G + \eta_{2G}(n)] + \eta_{1G}(n)\}$$

$$\circledast h_{ps}(n) - \eta_{3G}(n) - \mu_{\alpha_B}w_B + \mu_{\alpha_G}w_G\bigg)^2\bigg\}$$

From the mutual independence and zero-mean assumption for $\eta_1(n)$, $\eta_2(n)$ and $\eta_3(n)$, the crossing products $E\{\eta_k(n)\eta_l(n)\}$, $k \neq l$ are disregarded, and equation above results:

$$\sigma_{d_b}^2 = \frac{1}{N^2}\sum_{n=1}^{N}\sum_{m=1}^{N}E\{\alpha_B^2[\eta_{2B}(n)\circledast h(n)][\eta_{2B}(m)\circledast h(m)]\}$$

$$+ \frac{1}{N^2}\sum_{n=1}^{N}\sum_{m=1}^{N}E\{[\eta_{1B}(n)\circledast h(n)][\eta_{1B}(m)\circledast h(m)]\}$$

$$+ \frac{1}{N^2}\sum_{n=1}^{N}\sum_{m=1}^{N}E\{\eta_{3B}(n)\eta_{3B}(m)\}$$

$$+ \frac{1}{N^2}\sum_{n=1}^{N}\sum_{m=1}^{N}E\{\alpha_G^2[\eta_{2G}(n)\circledast h(n)][\eta_{2G}(m)\circledast h(m)]\} \qquad (3.97)$$

$$+ \frac{1}{N^2}\sum_{n=1}^{N}\sum_{m=1}^{N}E\{[\eta_{1G}(n)\circledast h(n)][\eta_{1G}(m)\circledast h(m)]\}$$

$$+ \frac{1}{N^2}\sum_{n=1}^{N}\sum_{m=1}^{N}E\{\eta_{3G}(n)\eta_{3G}(m)\}$$

$$+ E\{[w_B(\alpha_B - \mu_{\alpha_B})]^2\} + E\{[w_G(\alpha_G - \mu_{\alpha_G})]^2\}.$$

Let $z_{1B}(n) = \eta_{1B}(n) \circledast h(n)$ and $z_{2B}(n) = \eta_{2B}(n) \circledast h(n)$, thus:

$$\sigma_{d_b}^2 = \frac{1}{N^2}\sum_{n=1}^{N}\sum_{m=1}^{N}E\{\alpha_B^2 z_{2B}(n)z_{2B}(m)\} + \frac{1}{N^2}\sum_{n=1}^{N}\sum_{m=1}^{N}E\{z_{1B}(n)z_{1B}(m)\} + \frac{\sigma_{\eta_{3B}}^2}{N}$$

$$+ \frac{1}{N^2}\sum_{n=1}^{N}\sum_{m=1}^{N}E\{\alpha_G^2 z_{2G}(n)z_{2G}(m)\} + \frac{1}{N^2}\sum_{n=1}^{N}\sum_{m=1}^{N}E\{z_{1G}(n)z_{1G}(m)\} + \frac{\sigma_{\eta_{3G}}^2}{N}$$

$$+ w_B^2\sigma_{\alpha_B}^2 + w_G^2\sigma_{\alpha_G}^2$$

$$= \frac{1}{N^2}\sum_{n=1}^{N}\sum_{m=1}^{N}R_{z_{2G}}(m,n)(\sigma_{\alpha_B}^2 + \mu_{\alpha_B}^2) + \frac{1}{N^2}\sum_{n=1}^{N}\sum_{m=1}^{N}R_{z_{1B}}(m,n) + \frac{\sigma_{\eta_{3B}}^2}{N}$$

$$+\frac{1}{N^2}\sum_{n=1}^{N}\sum_{m=1}^{N}R_{z_{2G}}(m,n)(\sigma_{\alpha_G}^2+\mu_{\alpha_G}^2)+\frac{1}{N^2}\sum_{n=1}^{N}\sum_{m=1}^{N}R_{z_{1G}}(m,n)+\frac{\sigma_{\eta_{3G}}^2}{N}$$

$$+w_B^2\sigma_{\alpha_B}^2+w_G^2\sigma_{\alpha_G}^2 \tag{3.98}$$

where $R_{z_{1B}}(m,n)$ and $R_{z_{2B}}(m,n)$ are the autocorrelation functions at the blurring filter output, for the input signals η_{1B} and η_{2B}, respectively. Let $R_{z_{1B}}(m,n)=r_h(m,n)\circledast r_{\eta_{1B}}(m,n)$, by observing the output properties of a linear system with random input. The variables $r_{\eta_{1B}}$ and r_h represent the autocorrelation functions of η_{1B} and of the impulse response of h, respectively. Therefore,

$$\sigma_{d_b}^2=\frac{\sigma_{\alpha_B}^2+\mu_{\alpha_B}^2}{N^2}\sum_{n=1}^{N}\sum_{m=1}^{N}r_h(m,n)\circledast r_{\eta_{2B}}(m,n)$$

$$+\frac{1}{N^2}\sum_{n=1}^{N}\sum_{m=1}^{N}r_h(m,n)\circledast r_{\eta_{1B}}(m,n)+\frac{\sigma_{\eta_{3B}}^2}{N}$$

$$+\frac{\sigma_{\alpha_G}^2+\mu_{\alpha_G}^2}{N^2}\sum_{n=1}^{N}\sum_{m=1}^{N}r_h(m,n)\circledast r_{\eta_{2G}}(m,n) \tag{3.99}$$

$$+\frac{1}{N^2}\sum_{n=1}^{N}\sum_{m=1}^{N}r_h(m,n)\circledast r_{\eta_{1G}}(m,n)+\frac{\sigma_{\eta_{3G}}^2}{N}+$$

$$+w_B^2\sigma_{\alpha_B}^2+w_G^2\sigma_{\alpha_G}^2$$

Since η_{1B} and η_{2B} are modeled as uncorrelated noise, $r_{\eta_{1B}}(m,n)$ and $r_{\eta_{2B}}(m,n)$ are represented by an impulse at the origin with amplitude $\sigma_{\eta_{1B}}^2$ and $\sigma_{\eta_{2B}}^2$, respectively. Since $\sum_n h(n)=1$, $\sum_{m,n}r_h(m,n)=1$, and (3.99) becomes

$$\sigma_{d_b}^2=\frac{(\sigma_{\alpha_B}^2+\mu_{\alpha_B}^2)\sigma_{\eta_{2B}}^2+\sigma_{\eta_{1B}}^2+\sigma_{\eta_{3B}}^2}{N}+\frac{(\sigma_{\alpha_G}^2+\mu_{\alpha_G}^2)\sigma_{\eta_{2G}}^2+\sigma_{\eta_{1G}}^2+\sigma_{\eta_{3G}}^2}{N}$$

$$+w_B^2\sigma_{\alpha_B}^2+w_G^2\sigma_{\alpha_G}^2, \tag{3.100}$$

where $\sigma_{\eta_{2B}}^2=(w_B-w_B^2)$.

First, assume that the detection is performed using only the blue channel, for example. Considering the $S=2$ (or 1 bit) case, for example, the conditional error probability p_{01} given that bit 1 was transmitted is described by $p_{01}=\Pr(d_b>\lambda_B|\text{bit}=1)$, where λ_b is a decision threshold. Using the complementary error function erfc, $p_{01}=\frac{1}{2}\text{erfc}\left(\frac{\lambda_b-\mu_{d_b/1}}{\sqrt{2\sigma_{d_b/1}^2}}\right)$, where $\mu_{d_b/1}$ and $\sigma_{d_b/1}^2$ are respectively the mean and the variance of d_b for bit 1. Equivalently, if bit 0 is transmitted, the conditional error probability is given by $p_{10}=\frac{1}{2}\text{erfc}\left(\frac{\mu_{d_b/0}-\lambda_b}{\sqrt{2\sigma_{d_b/0}^2}}\right)$, where $\mu_{d_b/0}$ and

$\sigma^2_{d_b/0}$ are respectively the mean and the variance of d_b for bit 0. Finally, the average error probability is expressed by

$$P_{e_{d_b}} = P_0 p_{10} + P_1 p_{01} \tag{3.101}$$

where P_0 and P_1 are the probabilities of occurrence of bits 0 and 1, respectively. A similar approach is valid for the red channel.

3.7.3.2 Bhattacharyya Bound on the Error Rate

To reduce the error rate, metrics d_b and d_r of the blue and red channels described in (3.92) are combined into a single metric. For this task, the Bayes decision rule is employed. Because the metrics are assumed to be normally distributed, the Bayes decision rule guarantees the lowest average error rate [23].

It is relatively straightforward to determine the error rate using the metrics independently (as given in (3.101)). However, the error rate for the Gaussian case when the metrics are combined with the Bayes classifier are not easily computable, specially in high dimensions. However, in the two-category ($S = 2$) case, an upper bound on the error rate is given by the Bhattacharyya bound for Gaussian variables [23]. A closer bound is given by the Chernoff bound [23], however the Bhattacharyya bound is used in this work to avoid determination of β, (required for Chernoff), as discussed in [23].

Therefore, the classification error probability is:

$$P_{e_d} = \sqrt{P_0 P_1} e^{-k} \tag{3.102}$$

where k is given by

$$k = \frac{1}{8}(\mu_{d_{/1}} - \mu_{d_{/0}})^T \left(\frac{\Sigma_{/0} + \Sigma_{/1}}{2} \right)^{-1} (\mu_{d_{/1}} - \mu_{d_{/0}}) + \frac{1}{2} \ln \frac{\left| \frac{\Sigma_{/0} + \Sigma_{/1}}{2} \right|}{\sqrt{|\Sigma_{/0}||\Sigma_{/1}|}} \tag{3.103}$$

where $\Sigma_{/0}$ and $\Sigma_{/1}$ represent the covariance matrices corresponding to bit 0 and bit 1, respectively.

Notice that k depends on the results from (3.95) and (3.100). Applying (3.103) to (3.102), the plot presented in Fig. 3.33 is generated. This figure illustrates the upper bound on the error rate given in (3.102) as a function of the modulation level in the red and blue channels, given by w_R and w_B, respectively. The noise parameters for generating this plot are given in the experiments section.

Notice that modulating both color channels yields a lower error rate than modulating only one color channel, at the expense of increased perceptual impact. The perceptual impact caused by the color modulation is discussed in the next section.

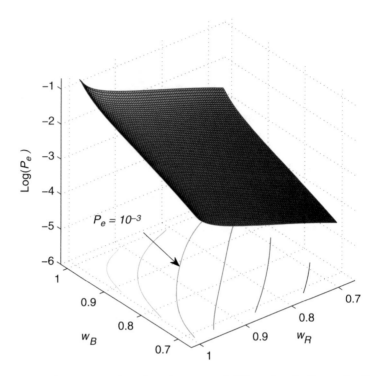

Fig. 3.33 Bhattacharyya upper bound on the error probability using the Bayes classifier, as a function of the modulation in the red and blue channels, ©2009 IEEE. Reprinted from [13] with permission license no. 4220870332019

3.7.4   Perceptual Evaluation of Color Modulation

Several metrics that try to evaluate the visibility of color variations in uniform targets have been proposed in the literature [43, 66]. The CIELAB metric [65] is a standard that describes an adequate transformation of physical image measurements into perceptual differences. The metric has been applied in industry for several years, and it is accounted as an acceptable tool for measuring perceptual difference between large uniform patches of colors [65].

Although other color difference metrics can be used, using the CIELAB color difference metric discussed in [65], it is possible to estimate the perceptual impact of the modulation level, as the original image color is black (black characters). The perceptual distortion metric D is given by:

$$D = \sqrt{(L_0 - L)^2 + (a_0 - a)^2 + (b_0 - b)^2} \tag{3.104}$$

where L_0, a_0 and b_0 represent the parameters of the original color in the $L^*a^*b^*$ color space.

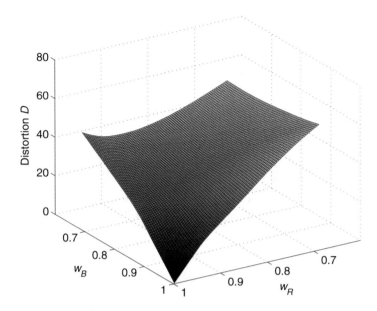

Fig. 3.34 Perceptual impact based on the CIELAB color space caused by the character color modulation, as a function of the modulation in the red and blue channels, ©2009 IEEE. Reprinted from [13] with permission license no. 4220870332019

The graph in Fig. 3.34 shows the perceptual difference D between black and a color generated by the modulation of the blue and red channels. In this figure, the green channel is not modulated.

The detection error probability and the perceptual impact as a function of the modulation in the red and blue channels were derived in Sects. 3.7.3 and 3.7.4, respectively. With this information, it is possible to determine the trade-off between these two conflicting properties, and use the optima modulation values for a given error rate or perceptual impact. Considering the least perceptual modulation for a given error rate, for example, the following optimization is performed

$$\underset{w_B, w_R \in [0,1] | P_{e_d} = P'}{\text{argmin}} D(w_B, w_R) \tag{3.105}$$

which returns the values w_B and w_R for which the distortion D is minimum, constrained to an error probability equal to $P_{e_d} = P'$. This optimization process cannot be performed analytically. Instead, the modulation values which correspond to P' are tested regarding the distortion given in (3.104), where the minimum value is chosen as optimum.

Figure 3.35 illustrates the distortion curve as a function of the modulation w_B and w_R, for an error probability upper bound equal to $P_{e_d} = 0.0001$, based on the plot

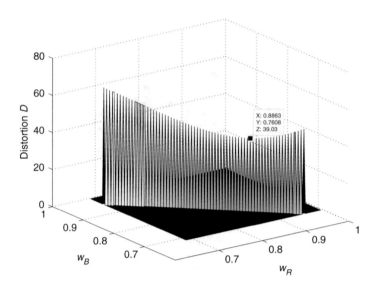

Fig. 3.35 Illustration of the distortion as a function of w_B and w_R, considering an error probability $P_e = 0.0001$, ©2009 IEEE. Reprinted from [13] with permission license no. 4220870332019

given in Fig. 3.33. Notice that when the distortion parameters of PS channel used are obtained, the error rates are obtained via (3.103) and the modulation levels that cause the least distortion (according to (3.104)) can be employed.

3.7.5   Auxiliary Metric to Improve Detection

As discussed in Sect. 3.1.3, in order to reduce the Moiré effect, the dithering matrices for each channel are rotated to be as far apart in angle as possible. This rotation is usually set to 0° for the Y screen, 15° for the M screen, 45° for the K screen, and 75° for the C screen. Therefore, when a character suffers color modulation, an "orientation modulation" also occurs as a consequence. This section proposes to use this characteristic an auxiliary metric to help improving the detection in TCM.

Several methods for texture orientation detection have been proposed in the literature [36, 47], such as Fourier methods, Gabor filters [27] and Tamura features [54]. In this application, however, the expected screen angle is known, and the use of a directional edge detection filter 'matched in angle' is a simple and efficient solution. The average energy of the filtered version of the character is an efficient discrimination metric to determine the predominant screen angle, therefore determining whether or not the character is color modulated. The maximum energy is attained when the filtering angle is set to 90° to the predominant screen angle of the modulated character. The orientation detection metric is given by:

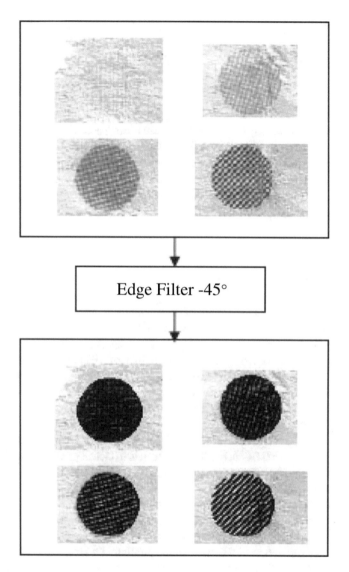

Fig. 3.36 Halftone color patterns filtered with $-45°$ directional edge filter, ©2009 IEEE. Reprinted from [13] with permission license no. 4220870332019

$$d_a = \frac{1}{N} \sum_{n=1}^{N} y_f(n) \qquad (3.106)$$

where $y_f = y \circledast f_{\text{dir}}$ and f_{dir} is a directional edge detector filter.

As an example, which is visible to illustrate the underlying process, Fig. 3.36 illustrates a version of Fig. 3.30 filtered with a edge detector filter f_{dir} set to a $-45°$ angle. Notice that the filtered patterns energies vary depending on the color used.

Although not as robust as the metrics described in Sect. 3.7.3, this approach has classification power and reduces the classification error rate, when combined with the other metrics, such that

$$\mathbf{d}_3 = [d_b \quad d_r \quad d_a]^\top \tag{3.107}$$

where the metrics are combined according to the Bayes classifier. Although in this three-feature case the Bhattacharyya bound discussed in Sect. 3.7.3.2 cannot be applied, the experiments in Sect. 3.7.6 illustrate the results of using the metric described in (3.107) work as an efficient detection metric.

3.7.6 Experiments

3.7.6.1 Blue Channel Modulation

Consider the 1 bit/element case ($S = 2$). A large sequence of $K = 30,360$ characters (as in 'abcdef...') is printed, with font type 'Arial', size 12 points.

Prior to printing, the blue channel of the character sequence was modulated with a gain $w_{B_i} = 1$ (no color alteration) for odd i, $i = 1, 3, \ldots, K - 1$, and with a gain $w_{B_i} = 0.81$ for even i, $i = 2, 4, \ldots, K$. Using these values, empirical tests indicate that it is hard for a human observer to distinguish between a modulated and a non-modulated character. In this case, according to the perceptual distortion metric described in (3.104), the distortion is given by $D = 34.67$. In fact, the empirical tests performed with human observers have illustrated that $D < 40$ are usually not perceivable by the human eye.

The elements with no color alteration ($w_{B_i} = 1$) carry bit 0, and the elements modulated with $w_{B_i} = 0.81$ carry bit 1. The task is to classify each printed character as having a bit 0 or bit 1 embedded into it. In this example, the difference $d_b = d_B - d_G$ is used as a detection metric, as given in (3.93).

The histogram of the detection results are presented in Fig. 3.37. The error rates for this experiment are presented in Table 3.1, in the row corresponding to d_b. For comparison, this table also presents error rates using simple TLM, where the three channels are modulated with the same gain, simulating a single-channel gray level modification only. With the same equipment and resolutions, the modulation gains for TLM was 0.81, which causes $D = 39.59$, also making the modulation hard to perceive. Notice that color modulation presents a lower error rate in comparison with simple luminance modulation.

3.7.6.2 Red Channel Modulation

This experiment is similar to the previous, however the red channel is modulated, with a gain $w_{R_i} = 0.85$ for $i = 2, 4, \ldots, K$. This gain value was modified such that the distortion D remained the same as in the blue channel case, that is, $D =$

Fig. 3.37 Histogram of the detection metric d_B, for the blue channel, ©2009 IEEE. Reprinted from [13] with permission license no. 4220870332019

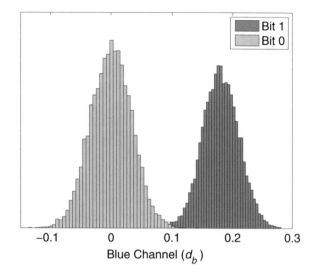

Blue Channel (d_b)

Table 3.1 Experimental error rates for TCM and TLM

Detection type	Number of errors	Error rate
TLM	139	4.6×10^{-3}
Blue modulation (d_b)	35	1.2×10^{-3}
Red modulation (d_r)	53	1.7×10^{-3}
Red-blue modulation ($\mathbf{d_2}$)	6	1.96×10^{-4}
Orientation (d_a)	1040	3.39×10^{-2}
Red-blue-orientation ($\mathbf{d_3}$)	2	6.59×10^{-5}

34.67. In this example, the difference $d_r = d_R - d_G$ is used as a detection metric, as described in (3.93). The histogram of the detection results and the error rates for this experiment are given in Fig. 3.38 and Table 3.1, respectively. Notice that when no modulation is performed (bit 0), no difference between the averages of the red and green channels is expected, in agreement with the experimental results.

3.7.6.3 Combined Blue-Red Channel Modulation

In this experiment, the blue and the red channels are modulated in characters c_i, $i = 2, 4, \ldots, K$. The modulation gain is given by $w_B = 0.76$ and $w_R = 0.88$, which are the modulation values that yield the minimum distortion for an error probability equal to 0.0001, according to the result presented in Fig. 3.35. In this case, the distortion of the modulated characters is given by $D = 39.03$.

In the detection, the information from the blue and red channels are combined according to the Bayes classifier, as described in Sect. 3.7.3.2. Figure 3.39 shows a scattered plot with the decision function obtained for this set of tests.

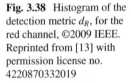

Fig. 3.38 Histogram of the detection metric d_R, for the red channel, ©2009 IEEE. Reprinted from [13] with permission license no. 4220870332019

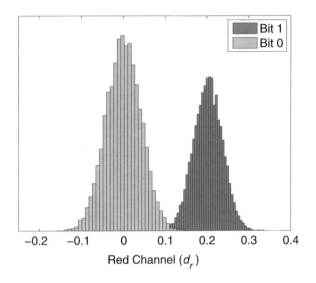

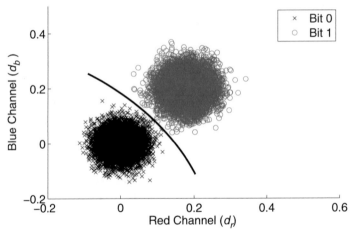

Fig. 3.39 Scatter plot illustrating curve separating the detection values for bit 1 and bit 0, ©2009 IEEE. Reprinted from [13] with permission license no. 4220870332019

The experimental error rates combining the two channels with the Bayes classifier is given in Table 3.1, in the row corresponding to $\mathbf{d_2}$. This illustrates that the detection using two color channels is more efficient than single channel detection, at the expense of a slight increase in the distortion. However, this distortion is still hardly perceived by a human observer.

3.7.6.4 Auxiliary Angular Detection

This experiment presents the improved performance using the approach described in Sect. 3.7.5, where the average energy d_a of the output of the directional filter f_{dir}

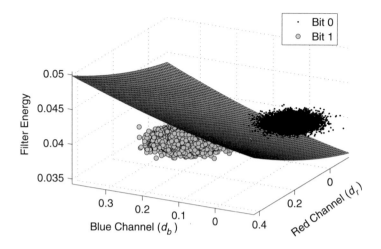

Fig. 3.40 Plot illustrating surface separating the detection values for bit 1 and bit 0, using d_b, d_r and d_a as detection metrics, ©2009 IEEE. Reprinted from [13] with permission license no. 4220870332019

is used as a detection metric. The experiments illustrate that when d_a is used alone, it does not bring an error rate as low as the metrics d_b and d_r, as shown in Table 3.1. However, d_a can be combined with d_b and d_r used in Experiments 1 and 2 using the Bayesian approach, to help reducing the error rate. Results using this approach are also given in Table 3.1, in the row corresponding to $\mathbf{d_3}$.

Figure 3.40 shows a 3-D plot showing a surface separating the modulated from the non-modulated characters. The directional filter used in this example was oriented at 165°, which corresponds to the blue screen angle plus 90°.

3.7.6.5   Variable Modulation Gains

This experiment observes the error rates for different gains in the red-blue channel modulation case. Instead of testing only for $w_{B_i} = 0.76$ and $w_{R_i} = 0.88$, $i = 2, 4, \ldots, K$ (even i), the tested gains are $w_{R_i} = w_{B_i} = 0.97$, $w_{R_i} = w_{B_i} = 0.9$, $w_{R_i} = w_{B_i} = 0.84$, for even i. The goal of this experiment is to evaluate how accurate is the upper bound on the error rate compared to practical results.

Figure 3.41 shows a plot with the theoretical (full line) upper bound on the error probability described by (3.103) derived in the analysis of Sect. 3.7.3, as a function of the gain $w_R = w_B$. Because $w_R = w_B$, a two-dimensional plot is used to illustrate the errors, instead of the 3-D plot shown in Fig. 3.33. Figure 3.41 also shows the experimental error rates (represented by crosses) as a function of the difference between the original character luminance and the gain $w_{B,R}$. This figure indicates the similarity between the theoretical and the experimental results after printing and scanning.

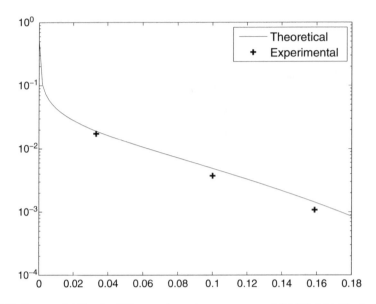

Fig. 3.41 Error probability for different gains, using $w_B = w_R = 0.97, 0.9, 0.84$. The horizontal axis indicates the luminance difference between the original and the modulated character, for the red and blue channels. The full line represents the theoretical error rate derived in Sect. 3.7.3, the cross dots represent the experimental error rates after PS, ©2009 IEEE. Reprinted from [13] with permission license no. 4220870332019

3.8   Conclusions

This chapter has discussed text modulation as an approach to convey hidden information over hardcopy channels and also extended the TLM concept by using color as a modifiable feature, presenting analyses and experiments. These techniques are proven to be effective as their robustness and capacity have been estimated by modeling distortions that may arise in practical scenarios.

The next chapter discusses the overt approach (visible marks) to convey information also known as color barcode modulation, where the main concern is the density of information for a given robustness, namely, the capacity of the technique is important while the resulting distortion is not an issue provided that the designed paper code uses a small and confined area in the document.

References

1. A.M. Alattar, O.M. Alattar, Watermarking electronic text documents containing justified paragraphs and irregular line spacing. Proc. SPIE **5306**, 685–695 (2004)
2. T. Amano, A feature calibration method for watermarking of document images, in *IEEE Proceedings of the Fifth International Conference on Document Analysis and Recognition, ICDAR'99*, 20–22 Sept. 1999

3. R. Baitello, M. Barni, F. Bartolini, V. Cappellini, From watermark detection to watermark decoding: a PPM approach. Signal Process. **81**, 1261–1271 (2001)
4. F.A. Baqai, J.H. Lee, A.U. Agar, J.P. Allebach, Digital color halftoning. IEEE Signal Process. Mag. **22**(1), 87–96 (2005)
5. M. Barni, F. Bartolini, *Watermarking Systems Engineering: Enabling Digital Assets Security and Other Applications* (Marcel Dekker, New York, 2004)
6. B.E. Bayer, An optimum method for two-level rendition of continuous tone pictures, in *IEEE International Conference on Communications*, Conference Records (1973), pp. 26-11–26-15
7. P.V.K. Borges, Text luminance modulation for hardcopy watermarking. Ph.D. thesis, Queen Mary, University of London, 2007
8. P.V.K. Borges, J. Mayer, Analysis of position based watermarking. Pattern Anal. Appl. **9**, 70–82 (2006)
9. P.V.K. Borges, J. Mayer, Text luminance modulation for hardcopy watermarking. Signal Process. **87**, 1754–1771 (2007)
10. P.V.K. Borges, E. Izquierdo, J. Mayer, A practical protocol for digital and printed document authentication, in *15th EURASIP European Signal Processing Conference, EUSIPCO 2007*, Poznan, 4–7 Sept. 2007
11. P.V.K. Borges, E. Izquierdo, J. Mayer, Efficient side information encoding for text hardcopy documents, in *IEEE Conference on Advanced Video and Signal Based Surveillance, 2007. AVSS 2007* (IEEE, New York, 2008)
12. P.V.K. Borges, J. Mayer, E. Izquierdo, Document image processing, for paper side communications. IEEE Trans. Multimedia **10**(7), 1277–1287 (2008)
13. P.V.K. Borges, J. Mayer, E. Izquierdo, Robust and transparent color modulation for text data hiding. IEEE Trans. Multimedia **10**(8), 1479–1489 (2008)
14. A. Bovik, *Handbook of Image & Video Processing* (Academic, Cambridge, 2000)
15. J. Brassil, S. Low, N.F. Maxemchuk, L. O'Gorman, Electronic marking and identification techniques to discourage document copying. IEEE J. Sel. Areas Commun. **13**, 1495–1504 (1995)
16. J.T. Brassil, S. Low, N.F. Maxemchuk, Copyright protection for the electronic distribution of text documents. Proc. IEEE **87**(7), 1181–1196 (1999)
17. Compris Intelligence - Germany, www.textmark.com
18. J. Cannons, P. Moulin, Design and statistical analysis of a hash-aided image watermarking system. IEEE Trans. Image Process. **13**(10), 1393–1408 (2004)
19. I.J. Cox, M.L. Miller, J.A. Bloom, *Digital Watermarking* (Morgan Kaufmann, Burlington, 2002)
20. M.C. Davey, D.J.C. MacKay, Reliable communication over channels with insertions, deletions, and substitutions. IEEE Trans. Inf. Theory **47**, 687–698 (2001)
21. N. Degara-Quintela, F. Pérez-González, Visible encryption: using paper as a secure channel, in *Security and Watermarking of Multimedia Contents V*, ed. by P.W. Wong, E.J. Delp. Proceedings of SPIE, Santa Clara, CA, January 2003, vol. 5020
22. F. Deguillaume, S. Voloshynovskiy, T. Pun, Character and vector graphics watermark for structured electronic documents security, September 2004. US Patent Application 10/949,318
23. R.O. Duda, P. Hart, D.G. Stork, *Pattern Classification*, 2nd edn. (Wiley, Hoboken, 2001)
24. H. Freeman, Computer processing of line-drawing images. ACM Comput. Surv. **6**(1), 57–97 (1974)
25. A. Glassner, Inside moire patterns. IEEE Comput. Graph. Appl. **17**(6), 97–101 (1997)
26. R.C. Gonzalez, R.E.Woods, *Digital Image Processing* (Addison-Wesley, Boston, 1992)
27. M. Haley, B.S. Manjunath, Rotation-invariant texture classification using a complete space-frequency model. IEEE Trans. Image Process. **8**, 255–269 (1999)
28. J. HÅstad, R. Impagliazzo, L.A. Levin, M. Luby, A pseudorandom generator from any one-way function. SIAM J. Comput. **28**(4), 1364–1396 (1999)
29. D. Huang, H. Yan, Interword distance changes represented by sine waves for watermarking text images. IEEE Trans. Circuits Syst. Video Technol. **11**, 1237–1245 (2001)

30. International Commission on Illumination (CIE), in *Recommendations on Uniform Color Spaces, Color Difference Equations, Psychometric Color Terms*. Publication CIE 15 (E.-1.3.1), Supplement No. 2 (Bureau Central de la CIE, Vienna, 1971)

31. S.H. Kim, J.P. Allebach, Impact of HVS models on model-based halftoning. IEEE Trans. Image Process. **11**(3), 258–269 (2002)

32. Y.-W. Kim, K.-A. Moon, I.-S. Oh, A text watermarking algorithm based on word classification and inter-word space statistics, in *IEEE Proceedings of the Seventh International Conference on Document Analysis and Recognition (ICDAR'03)*, 2003

33. X. Li, X. Xue, Fragile authentication watermark combined with image feature and public key cryptography, in *7th Int'l Conference on Signal Processing, ICSP'04*, 2004

34. C.-Y. Lin, S.-F. Chang, Distortion modeling and invariant extraction for digital image print-and-scan process, in *International Symposium on Multimedia Information Processing, ISMIP 99*, Taipei, 1999

35. C.-Y. Ling, Public watermarking surviving general scaling and cropping: an application for print-and-scan process, in *Multimedia and Security Workshop at ACM Multimedia 99*, Orlando, FL, October 1999

36. F. Liu, R.W. Picard, Periodicity, directionality, and randomness: wold features for image modeling and retrieval. IEEE Trans. Pattern Anal. Mach. Intell. **18**(7), 722–733 (1996)

37. P. Lennie, Color vision: putting it together. Curr. Biol. **10**(16), R589–R591 (2000)

38. S. Low, N.F. Maxemchuk, Performance comparison of two text marking methods. IEEE J. Sel. Areas Commun. **16**, 561–572 (1998)

39. S. Low, N.F. Maxemchuk, Capacity of text marking channel. IEEE Signal Process. Lett. **7**(12), 345–347 (2000)

40. S. Low, N.F. Maxemchuk, A.M. Lapone, Document identification for copyright protection using centroid detection. IEEE Trans. Commun. **46**(3), 372–383 (1998)

41. D.G. Manolakis, V.K. Ingle, S.M. Kogon, *Statistical and Adaptive Signal Processing* (McGraw-Hill, New York, 2000)

42. M. Mese, P.P. Vaidyanathan, Recent advances in digital halftoning and inverse halftoning methods. IEEE Trans. Circuits Syst. I: Fundam. Theory Appl. **49**(6), 790–805 (2002)

43. A. Mojsilovic, J. Hu, R. Safranek, Perceptually based color texture features and metrics for image retrieval, in *IEEE International Conference on Image Processing*, October 1999, vol. 3

44. J. Mulligan, Digital halftoning methods for selectively partitioning error into achromatic and chromatic channels. Proc. SPIE **1249**, 261–270 (1990)

45. M. Norris, E.H.B. Smith, Printer modeling for document imaging, in *Proceedings of the International Conference on Imaging Science, Systems and Technology, CISST'04*, Las Vegas, NV, June 21–24, 2004

46. N. Otsu, A threshold selection method from gray-level histograms. IEEE Trans. Syst. Man Cybern. **9**(1), 62–66 (1979)

47. N. Prins, F.A.A. Kingdom, Detection and discrimination of texture modulations defined by orientation, spatial frequency, and contrast. J. Opt. Soc. Am. **20**(3), 401–410 (2003)

48. J. Riordan, *Introduction to Combinatorial Analysis* (Dover Publications, Mineola, 2002)

49. B. Sanguinetti, G. Traverso, J. Lavoie, A. Martin, H. Zbinden, Perfectly secure steganography: hiding information in the quantum noise of a photograph. Phys. Rev. **93**(1), 012336 (2016)

50. G. Sharma, H. Trussel, Digital color imaging. IEEE Trans. Image Process. **6**(7), 901–932 (1997)

51. B. Sklar, *Digital Communications - Fundamentals and Applications*, 2nd edn. (Prentice-Hall, Upper Saddle River, 2001)

52. K. Solanki, U. Madhow, B.S. Manjunath, S. Chandrasekaran, Modeling the print-scan process for resilient data hiding, in *Proceeding of SPIE. Electronic Imaging*, vol. 5681 (2005)

53. R.L. Stevenson, Inverse halftoning via MAP estimation. IEEE Trans. Image Process. **6**(4), 574–583 (1997)

54. H. Tamura, S. Mori, T. Yamawaki, Texture features corresponding to visual perception. IEEE Trans. Syst. Man Cybern. **SMC-8**(6), 460–473 (1978)

55. R.A. Ulichney, *Digital Halftoning* (MIT Press, Cambridge, 1988)

56. R.A. Ulichney, Dithering with blue noise. Proc. IEEE **76**(1), 56–79 (1988)
57. R. Villán, S. Voloshynovskiy, O. Koval, T. Pun, Multilevel 2D bar codes: towards high capacity storage modules for multimedia security and management, in *Proceedings of SPIE Photonics West, Electronic Imaging 2005, Security, Steganography, and Watermarking of Multimedia Contents VII (EI120)*, San Jose, CA, 16–20 January 2005
58. R. Villán, S. Voloshynovskiy, O. Koval, J. Vila, E. Topak, F. Deguillaume, Y. Rytsar, T. Pun, Text data-hiding for digital and printed documents: theoretical and practical considerations, in *Proceedings of SPIE-IST Electronic Imaging*, San Jose, CA, 2006
59. R. Villan, S. Voloshynovskiy, O. Koval, F. Deguillaume, T. Pun, Tamper-proofing of electronic and printed text documents via robust hashing and data-hiding, in *Proceedings of SPIE-IST Electronic Imaging, Security, Steganography, and Watermarking of Multimedia Contents IX*, San Jose, CA, 2007
60. P. Vinicius, K. Borges, J. Mayer, Document watermarking via character luminance modulation, in *Proceedings of IEEE Int'l Conference on Acoustics, Speech and Signal Processing, ICASSP'06*, Toulouse, May 2006
61. S. Voloshynovskiy, O. Koval, F. Deguillaume, T. Pun, Visual communications with side information via distributed printing channels: extended multimedia and security perspectives, in *Proceedings of SPIE Photonics West, Electronic Imaging 2004. Multimedia Processing and Applications*, San Jose, CA, 18–22 January 2004
62. P.W. Wong, N. Memon, Secret and public key image watermarking schemes for image authentication and ownership verification. IEEE Trans. Image Process. **10**(10), 1593–1601 (2001)
63. M. Wu, B. Liu, Data hiding in binary image for authentication and annotation. IEEE Trans. Multimedia **6**(4), 528–538 (2004)
64. H. Yang, A.C. Kot, Text document authentication by integrating inter character and word spaces watermarking, in *Proceedings of the IEEE International Conference on Multimedia and Expo*, 2004
65. X. Zhang, B.A. Wandell, Color image fidelity metrics evaluated using image distortion maps. Signal Process. **70**, 201–214 (1998)
66. X. Zhang, J.E. Farrell, B.A. Wandell, Applications of a spatial extension to CIELAB. Proc. SPIE **3025**, 154–157 (1997)

Chapter 4
Print Codes

This chapter[1] discusses one type of overt (visible) communication by employing color print codes aiming to convey information over printed media. In the literature many print codes have been proposed for conveying information, ranging from simple shapes like the PARC (Palo Alto Research Center Incorporated) dataglyphs, the popular linear-1D and 2D barcodes [11] to the recent color barcodes or 3D barcodes such as the Microsoft High Capacity Color Barcode (HCCB). However, communication systems that require screen displays such as the 4D barcodes [7] are not considered.

The goal is to embed high-density messages into printed materials in order to enable interesting hardcopy document applications involving security, authentication, item-level tagging, consumer/product interaction and physical-electronic round tripping. In this chapter, robust and high capacity print codes are employed to maximize information payload in a given printed page area. The maximization is subject to robustness to channel errors which include distortions originated by the printing and scanning processes and also to usual degradations introduced by user manipulation of printed documents in regular offices. An approach is presented that includes statistical print-and-scan channel characterization followed by designing of robust segmentation using visual cues. An unsupervised Bayesian color classification with expectation-maximization algorithm is employed for parameters estimation of a mixture of Gaussians model. We also provide the design of error correction codes. Results are provided to illustrate the performance evaluated under real channel and distortions conditions, including smudges, pen scribing and drops of liquid over the paper. The study provides a technique with high payload and robustness to distortions resulting of regular office hardcopy document handling such as print-and-scan channel and user manipulation.

---

[1]©2009 IEEE. Few paragraphs in this chapter are reprinted from [9] with permission license no. 4221411265137.

© Springer International Publishing AG 2018
J. Mayer et al., *Fundamentals and Applications of Hardcopy Communication*,
https://doi.org/10.1007/978-3-319-74083-6_4

4.1 Introduction

The literature indicates a variety of works that investigate high-capacity two-dimensional color print patterns codes to address several novel applications that require relatively large amount of information to be transmitted through printed media which is subsequently scanned in order to decode the information. These codes are coined as "3D barcodes" since including colors adds a third dimension to the 2D patterns. Most applications require robustness to analogue channel distortions in the printing and scanning processes as well as to various external degradations created by user manipulation.

Prior works on printed codes are mostly restricted to 1D barcodes which have limited payload, usually only few bytes, as compared to 3D barcodes. The work in [18] provide evidences that two-dimensional printed code technologies can significantly improve information transmission using paper. Thus, investigating high-capacity printed barcodes can potentially create new value for prints by improving security, print repurposability and supplemental information transmission. Moreover, related research on barcodes aimed to be captured by cell phones or consumer digital cameras has been proposed in the literature [8, 12, 13] and that technology and techniques can be also applied to our research that involves acquisition by scanners. More recent studies as in [5, 15, 18] have exploited the use of 2D patterns to improve capacity and robustness. Therefore, the literature provides solid evidences that significant improvements in capacity and robustness can be further achieved by exploiting hardcopy color communication over 2D patterns for high payload applications.

In order to achieve robustness and high-capacity information transmission over the print-and-scan channel a variety of issues and degradations needs to be addressed. These include distortions from many sources: (a) Paper (texture) and electronic noise; (b) The optical blurring; (c) Geometric and mechanical disturbances; (d) Ink mixing, spreading, spills, creases and aging; (e) Document manipulation by the user, smudges and pen scribing. In order to address these issues in this chapter we investigate some techniques: novel error correction coding, statistical classification and robust segmentation techniques. Another challenge is to develop technologies that are robust enough for general purpose use over a variety of branding (printing) devices, changes in ink and media properties and a variety of detection (scanning) devices. The investigation of new print codes to convey information over printed media using 2D color codes aims to discover new and interesting applications. Works in the literature include applications for restoration of aged printed photos with the help of 2D color codes printed on the back of the photo [14], content authentication using barcodes confined to small areas, security printing with smart labels [16, 17], and deterrents for branded product counterfeiting. This chapter provides techniques to improve algorithms operating with 2D color patterns to achieve higher capacity and robustness for novel applications demanding high information density or payload: number of bytes transmitted within a printed area of the paper.

4.2 Transmitting Through Print-Scan Channel

In general, the information is encoded (cryptography and ECC), modulated and embedded into the designed print codes before transmitting through the print-scan channel as illustrated in Fig. 4.1.

The channel consists of two main components: the distortions and transformations originated from the print-and-scan channel (PS) and the external distortions applied on the printed media. The PS channel presents both linear and non-linear distortions: ink spreading and paper noise (texture and imperfections); electronic, optical and motion blurring generated by the scanner device and rotation due to manual placement of paper media on the scanbed. A detailed modeling of these distortions for monochromatic PS channel is provided in [3]. The external distortions are generated by regular user manipulation (as opposed to malicious attacks) which may result in coffee spills, pen/pencil scribing, smudging, fingerprints and also the degradation caused by ink and paper media aging. Some of these distortions using the HP C4280 inkjet printer/scanner device are illustrated in Figs. 4.2 and 4.3.

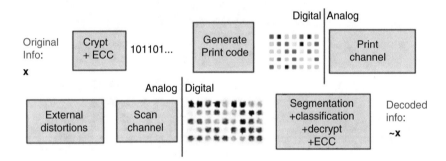

Fig. 4.1 Information encoding (cryptography and ECC), embedding, print-scan channel and information decoding system, ©2009 IEEE. Reprinted from [9] with permission license no. 4220850864216

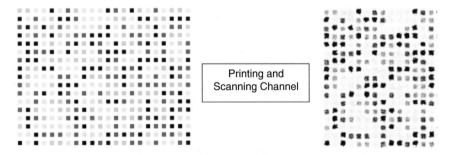

Fig. 4.2 Distortions in the PS channel, ©2009 IEEE. Reprinted from [9] with permission license no. 4220850864216

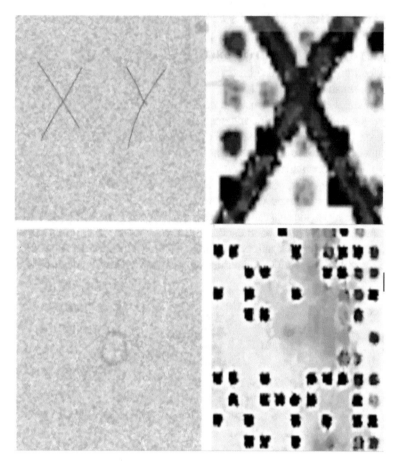

Fig. 4.3 External distortions due to pen scribing and coffee spills, ©2009 IEEE. Reprinted from [9] with permission license no. 4220850864216

4.3   The Print Codes Color Model

The scanning process acquires samples from the printed page image and delivers them as triplet {R,G,B} of values in the RGB color system. However, as most consumer printing devices employ the complement CMYK color model usually with four ink cartridge colors (cyan, magenta, yellow and black), it is natural and more convenient to represent the pixels in the CMY color system. The transformation from pixels in RGB color model to CMYK is straightforward, each pixel is represented as a 3-dimensional sample $x = [1 - R \ \ 1 - G \ \ 1 - B]^T = [C \ \ M \ \ Y]^T$.

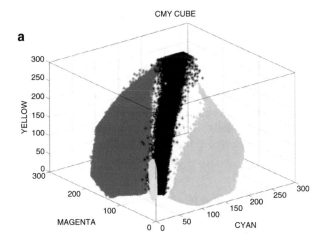

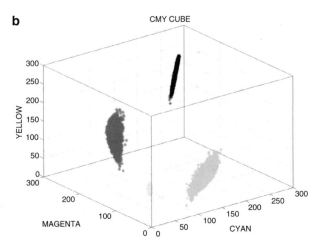

Fig. 4.4 (a) Individual samples color spreading. (b) Block sample means color spreading due to the PS channel, ©2009 IEEE. Reprinted from [9] with permission license no. 4220850864216

The actual printed codes are based on CMYK ink cartridge and the printed colors naturally deviates from pure CMYK colors due ink imperfections and aging and due to the texture of the print media. Additionally, the optical and mechanical parts of the scanner introduce color deviations for each acquired sample. We illustrate the resulting color deviations of individual samples from pure CMYK colors in Fig. 4.4a. These deviations are originated by optical and mechanical blurring and by the electronic noise intrinsics to the scanner device. As the deviations depends not only in ink variations (degradation, spreading and aging) but mostly due to the scanner optical and mechanical imperfections, we observe similar deviation on both inkjets and laser printer (in a lesser extent) technologies.

In order to mitigate such deviations we designed the print codes pattern as a square block of pixels of a given pure color. Instead of using the color of individual samples we measure the block mean color which considerably reduces the deviation. Thus, the color classification task is improved by using as feature the sample mean of the pixels within a block pattern. We can observe a considerable reduction of color deviation for each print code block as illustrated in Fig. 4.4b. While the design of print codes represented as a square block of a single color (either C, M, Y or K) reduces the color deviations and facilitates the classification (decoding), the payload is reduced inversely proportional to the size of the blocks.

4.4   Statistical Block Classification

The proposed technique requires to first classify the color of the printed codes and later on to decode the embedded information. We employ a color classification based on the Bayesian approach [1]. Each printed and scanned block pattern is classified in one of the C, M, Y or K classes. The classifier is based on a feature vector $y_j = f(x \in B_j)$ which is extracted from the acquired 3-dimensional samples (RGB sensor) within the j-th block pattern B_j. The vector y_j is a D-dimensional vector defined as a function of the samples within B_j. From the data we may estimate the class conditional probability density function (pdf), $p(y|\omega_i)$, of the feature vectors y belonging to class ω_i, where ω_1: cyan, ω_2: magenta, ω_3: yellow, ω_4: black. Moreover, the prior probability, $P(\omega_i)$, $i = 1, \ldots, 4$ may be also found from the data available. The most probable class of a given block sample, y_j, can be find as the posterior class probabilities [1]:

$$P(\omega_i|y) = \frac{p(y|\omega_i)P(\omega_i)}{\sum_{j=1}^{4} p(y|\omega_j)P(\omega_j)}. \tag{4.1}$$

Thus, the posteriors provide a statistical estimate of the embedded color in a given pattern block. We propose to build the feature vector using the sample mean and the average hue color (AVGH) [6] of the samples in the block B_j. The feature results in a 4-dimensional vector $y = [\frac{1}{M}\sum_{j=1}^{M} x_j; \ AVGH]$. Since the elements in y are sample averages of a certain number of samples, M, proportional to the size of the block, assuming that M is large, the Central Limit Theorem [1] can be applied to approximate the conditional pdfs of y as Gaussian. In this case, the desired class conditional probability density function for a given class ω_i can be modeled as:

$$p(y|\omega_i) = \frac{1}{(2\pi)^{3/2}|C_i|^{1/2}} \exp\left\{-\frac{1}{2}(y - \mu_i)^T C_i^{-1}(y - \mu_i)\right\} \tag{4.2}$$

The parameters of the class conditional, the covariance matrices, C_i, the mean vectors, μ_i and the prior probabilities $P(\omega_i)$, $i = 1, \ldots, 4$ can to be estimated given the observed data acquired from the print codes. The parameter estimation is addressed in the next section by modeling the received printed codes as a mixture of independent Gaussians.

4.5 Estimating the Parameters

At this point the designer has a choice: the density parameters can be estimated by either using supervised or unsupervised training. In the first case, the supervised training would require special reference blocks with previously defined known colors or accessing the original printing device. This is a faster and simpler approach but it requires some dedicated area in the print code, resulting in a smaller payload. On the other hand, the choice of this section, the unsupervised estimation approach does not require reference blocks or accessing the device. However, it requires some computational complexity to learn the parameters as explained in the follows.

Assuming that M is large enough such that the data is distributed according to a mixture of K Gaussians,

$$p(y) = \sum_{k=1}^{K} P(\omega_k) \frac{1}{(2\pi)^{D/2} |C_k|^{\frac{1}{2}}} e^{-\frac{1}{2}(y-\mu_k)^T C_k^{-1}(y-\mu_k)} \qquad (4.3)$$

the task is to estimate the parameters for the mixture of $K = 4$ Gaussians (four colors or classes). As the chosen feature vector has four descriptors, the dimension is $D = 4$. In this vector, the first three components are the average sample means inside of a block, namely, average of the cyan, magenta and yellow components. The fourth component is the average color hue of the block samples where color hue represents the angle in the HSV color model [6]. These descriptors have been chosen based on performance tests that indicated that they have a good discrimination property when compared to other options of descriptors. The three average descriptors help to reduce the color spreading as shown before, and the hue descriptor helps to discriminate among similar colors in the CMY cube space.

The unsupervised learning is achieved by employing the **Expectation-Maximization** (EM) algorithm [1] to estimate the parameters of (4.3). The algorithm is straightforward: The first phase is to estimate the mean vector parameter using the K-means clustering and then to estimate the initial covariance matrices using the closest patterns to the estimated mean vector. The second phase is iterative, it iterates between evaluating the expectation of y for each Gaussian density (statistical centroids for each class) and the maximization of new parameters (means and covariances) considering the computed expectations. The algorithm stops after the variations on the estimated parameters becomes smaller than a predefined value.

After learning the parameters, the classification task is simple: the four posteriors are computed using the estimated parameters. The color pixels embedded at block B_j are classified in the class i that results in the greatest $p(\omega_i|y)$. In the absence of user manipulations of the media that may induces degradations, the unsupervised estimation with EM algorithm and the Bayesian classification provides an optimal framework to address decoding under print and scan channel disturbances, provided that M is large enough. The next section Error Correction Codes (ECC) are proposed to address external noises, such as user manipulations, that may not follow the assumed mixture of Gaussian distribution for the received print codes.

4.6   Error Correction Codes

Some external distortions do not follow a regular probability density distribution such as Gaussian pdf. Examples of such distortions include coffee spills and pen scribing. These types of distortions need to be addressed with an error correction algorithm at encoding phase. The proper tradeoff between code robustness and code rate depends on the amount and type of external errors expected by the intended applications and uses of the print codes.

In order to design an ECC, we need to setup the print codes by defining the specific dimensions. We propose to use print codes with size of about one quarter (4 in by 4 in area) of document page. Each block has $M = 25$ and designed with a size of 5 by 5 pixels square block using a unique color (C, M, Y or K). These block are separated by gaps of 5 white pixels. This setup is not optimal for all applications and need to be properly defined upon the application requirements. By observing some experiments with pen scribing errors, we assume that typical external errors are confined within about 0.25% of the print codes area. With this specifications and dimensions chosen, we design the error correction codes to achieve a robustness to spill coffee and pen scribing distortions with area no larger than 0.5% of the print codes area and define a code redundancy of less than 20% of the message, resulting in a code rate larger than 0.82. We designed the standard Bose, Chaudhuri, and Hocquenghem (BCH) and the Reed-Solomon (RS) codes with hard decision [4] to achieve the aforementioned specifications.

As result, the print codes consist of a square of 240 by 240 colored square blocks as illustrated in the figures. It corresponds to a sequence of 240 by $240 = 57,600$ blocks $\rightarrow 57600 \times 2$ bits per colored block $= 115,200$ bits. The BCH codeword length is $n = 511$ bits, information word length is $k = 421$ bits, and a correction capacity of $t = 10$ bits per word is chosen. A bit "0" is appended at the end of each codeword, making the codeword length a power of 2. The print codes consists of a sequence of $115200/512 = 225$ binary codewords, comprising $225 \times 421 = 94,725$ information bits. For the RS code a similar approach is taken resulting in a higher correction capacity using $n = 511$, $k = 423$, $t = 44$ and $115200/512 \times 423 = 95,175$ information bits per print codes using up 4 in by 4 in area.

4.7   Segmentation of the Print Codes

The segmentation step implies in the localization of the color print codes within the document page. The overall system performance is highly dependent on this task [12]. The segmentation needs to be performed prior to the color classification and to the decoding using the ECC algorithm proposed. To proper segment the print codes the algorithm needs to address tough conditions such as non-uniform illumination of the scanning device, geometric distortions introduced by printing mechanics/heating and scanning alignment, proprietary scanner optical system and associated optical distortions, paper texture and imperfections, accidental user scribing and proprietary device ink properties.

The proposed segmentation approach initially relies on a robust estimation of the four corners of the print codes bounding box followed by creating a grid to separate individual blocks as illustrated by the lines in Fig. 4.5. To achieve such segmentation grid, the algorithm employs Otso's thresholding [10] operating on the saturation color component of the scanned image. It follows a sequence of morphological operators employed to estimate the bounding box of the print codes. The pixels along the edges of the bounding box are input to a least squares minimization algorithm, which estimate the four main lines of the bounding box, achieving reasonable robustness to boundary noise and finally producing an estimate of the four corners required to create the grid. The resulting grid has been shown to be robust to channel noise, internal scribing, coffee spills and geometric distortions up to a certain size of block, namely 10×10. When designing smaller blocks to achieve higher payload, auxiliary visual cues are employed in [12, 17] to address mechanical disturbances that causes misalignments and improve the localization of the print codes. The next section describes these visual cues employed to improve the segmentation of high density print codes to achieve higher payloads.

4.8   Addressing the Segmentation Issues in High Density Print Codes

In order to estimate the segmentation grid we employ a set of morphological operations after a linear regression of the boundary pixels step to determine the main four lines of the bounding box around the print codes. With these lines we achieve a robust estimation of the location of the four corners that generate the grid.

However, as the density of the print codes increases by using blocks smaller than 10 by 10 pixels, misalignments arise due to mechanical and optical issues in the printing and scanning devices. As a result, the internal grid becomes misaligned with the print codes, as illustrate in Fig. 4.7. The misalignment occurs mostly towards the center of the print codes since as near of the borders, due to the robust process of finding the corners, the grid is properly aligned. The

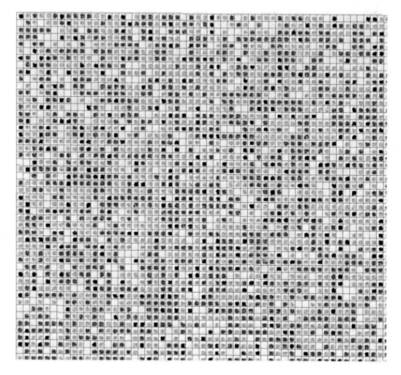

Fig. 4.5 Segmentation grid extracted from scanned blocks using size of 10 by 10 pixels, ©2009 IEEE. Reprinted from [9] with permission license no. 4220850864216

described process assures that the corners are properly detected provided that a linear regression is employed to mitigate the boundary noise. As a result we achieve a precise grid around the corners. However, it does not assure a proper alignment at center as optical and mechanical issues induce non uniform printing and scanning at the regions far from the corners. The misalignment is illustrate in Figs. 4.6 and 4.7.

The print codes classification is severely affected by these segmentation issues caused by the aforementioned misalignments as illustrated in Figs. 4.8 and 4.9. As a result, the EM algorithm requires more iterations trying to converge due to the mixing of block boundaries caused by faulty segmentation. In some cases, it requires over 200 iterations when usually only 10 iterations are required if proper alignment is in place. Although the convergence is achieved at very slow speed, the amount of misclassifications arise considerably. Thus, for higher print codes density, over 900 bytes/in², a different and more robust segmentation is needed. Some approaches based on region growing/splitting and gradient are not feasible as the amount of pixels in this problem, over 360,000 color pixels per squared inch, precludes the use of these interesting region/gradient based but computationally complex segmentation approaches.

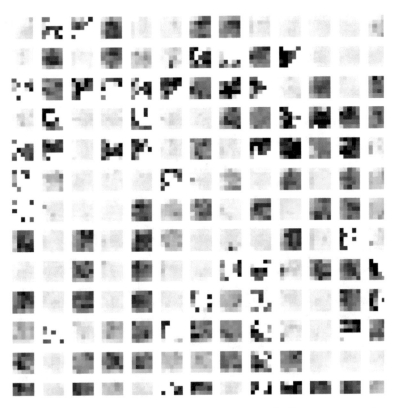

Fig. 4.6 Segmentation grid is properly defined at corners but misalignment exists at the center using size of 7 by 7 pixels

In order to address the segmentation for high densities print codes in a computational feasible way, we propose to include auxiliary visual cues to mitigate the misalignments at center. These visual cues are designed to be distributed around the print codes boundary. They are intended to align the grid at center and may also serve as non payload indicia (NPI), a way to improve classification or to use supervised classification instead of unsupervised as proposed. The distribution of the auxiliary visual cues is illustrated in Figs. 4.10 and 4.11. We find that as the proposed print codes are transmitted through the print-scan channel, the designed visual cues suffer the optical/mechanical distortions in the same way, as illustrated in Fig. 4.12. We notice that the black visual cues are more resilient to the channel due to the ink properties and they are employed as references to improve the segmentation grid. After the proper location of the boundary box and resulting four corners, the black visual cues are located to help to correct the grid and to achieve an improved segmentation as illustrated in Fig. 4.13.

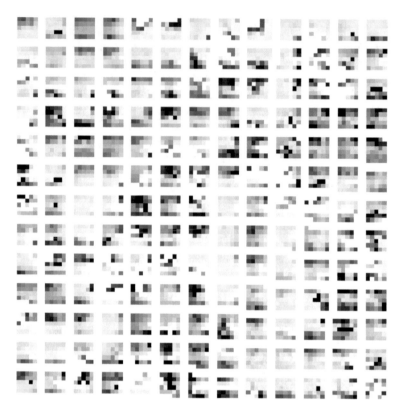

Fig. 4.7 Misalignment at the center of the segmentation grid extracted from scanned blocks using size of 7 by 7 pixels. Notice that the grid cross over the middle of the print codes

The grid segmentation is considerably improved for all regions of the print codes, as illustrated in Figs. 4.12 and 4.14. The proposed improved segmentation approach addresses high density print codes using block of 7 by 7 print codes, corresponding to 1836 bytes/in^2. The color estimation step in classification requires few iterations (as compared to 200 iterations without the additional visual cues) using the EM algorithm.

With the help of the visual cues, densities of 2500 bytes/in^2 (6 by 6 pixels in a block) are also achievable due to the improved grid alignment using printing and scanning at resolutions of 600 dpi. Further improvements are required to achieve a density over 3600 bytes/in^2 (5 by 5 pixels in a block) with great robustness. Since the M is small, the parameters estimation becomes less efficient. For these cases it is required to design proper ECC codes with the help of the visual cues and NPIs as well. Notice that the colors are predefined in these visual cues (NPIs), so they can be employed to provide parameter estimation and accurate Bayesian classification of the print codes.

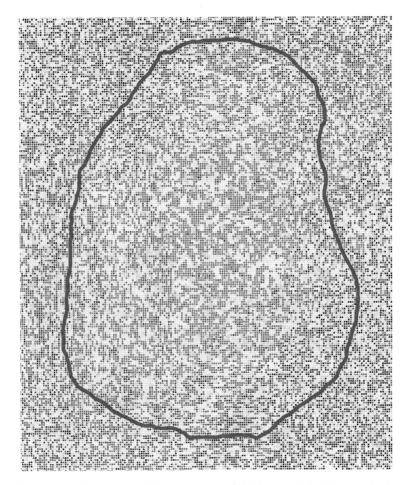

Fig. 4.8 Classification problems at the print codes center region resulting of the grid misalignment with blocks using size of 7 by 7 pixels

4.9 Experiments

In order to validate the performance of the print codes we select a a large set of inkjets and lasers printers (HPC4280, HPD6940, HPL1515, HPL2820) and scanners (HPC4280, HPL1120, HPS5590) to be used with A4 size and standard office paper media. The first set of experiments (using printer codes of 10 by 10 pixels) aims to observe the statistical properties of the printed and scanned data. Pages with blocks of one unique color are printed and the statistical moments (mean, variance, skewness and kurtosis) are extracted from each resulting color component from the observed printed and scanned pixels.

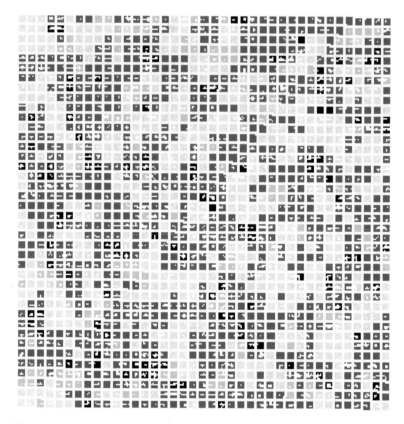

Fig. 4.9 Zooming color misclassifications at the print codes center region resulting of the grid misalignment with blocks using size of 7 by 7 pixels

Fig. 4.10 CMYK visual cues around the print codes boundary to address segmentation misalignment

The experiments indicate that the color statistics deviates slightly from a Gaussian distribution and the deviation is mostly dependent on the scanner optics. In these experiments, using the sample mean of the pixels inside of each square block (25 pixels), the kurtosis is kept around 20% of the value 3, and the skewness is less than 0.3. These values indicate that the distribution of the sample mean for the blocks can be approximated by a Gaussian pdf, where for perfect Gaussian pdf, the kurtosis would be 3 and skewness would be zero. For example, while printing

Fig. 4.11 Zooming on the proposed CMYK visual cues

magenta blocks, the scanned magenta component presented a kurtosis of 3.16 and skewness of −0.078. These values may vary considerably depending on the devices, the color printed and the component scanned.

The second set of experiments aims to observe the robustness of the proposed classifier. Initially, synthetic channel errors (typical noise, optical and geometrical disturbances) are created to resemble real printing channel errors. A similar approach was used in [2] to validate channel models and analysis. For these typical channel errors, the segmentation and classification algorithms perform sufficiently well such that all block colors are detected without errors. It is not necessary to introduce ECC to deal with these errors since the classifier provides very high robustness to them. External synthetic distortions along with synthetic channel errors are introduced to observe the robustness of the classifier for these distortions: global and local blurring, pen scribing and small cropping. The pen scribing and

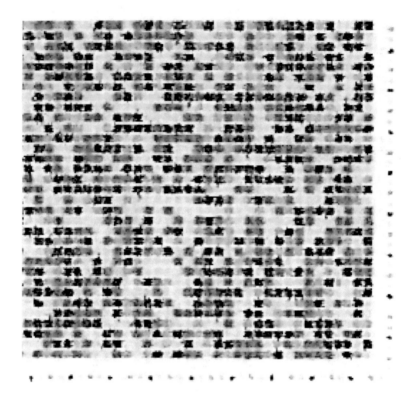

Fig. 4.12 Zooming on the proposed CMYK visual cues after PS channel

cropping distortions introduce classification errors such that ECC algorithms are needed to help to deal with these external errors.

The synthetic channel experiments indicate that the robustness of the ECC need to be proportional to the area of the external degradations on the document. Next, experiments with real channels are designed to introduce common external errors found in normal manipulation of documents, such as coffee spills, pen scribing and ink blurring due to fingerprints. The ECC algorithms are designed to sustain these external distortions. Figure 4.15 provides a few examples (from a large tested set) used to evaluate the system performance. The robust segmentation with the Bayesian classifier deals with the channel errors and the classifier takes the hard decision, while the ECC introduces at the print codes, before printing, the required redundancy to sustain the aforementioned external distortions.

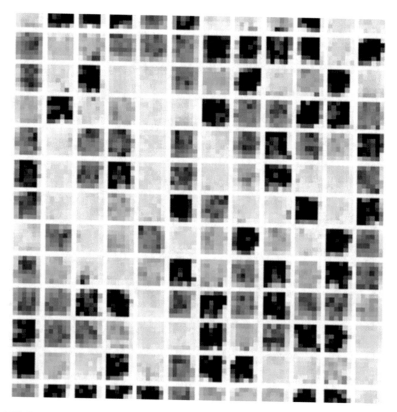

Fig. 4.13 Resulting improved segmentation grid at the print codes center

4.10 Conclusions

This chapter discuss a special type of overt communication of side information employing print codes and provides an approach for 3D barcode aiming high payload and robustness to the print and scan channel. The statistic framework is very efficient to address the problem at hand and the performance was confirmed by a large set of experiments.

The segmentation robustness and classification accuracy have been verified for a variety of devices. As illustrated in examples provided, the system robustly embeds at least 900 bytes/in^2 with 20% of redundancy due to the ECC, resulting in an information payload amount of around 750 bytes/in^2, which is adequate for the Photoplus [14] and other applications. The density can be increased by reducing the size of the blocks up to a certain limit imposed by segmentation issues and device limitations.

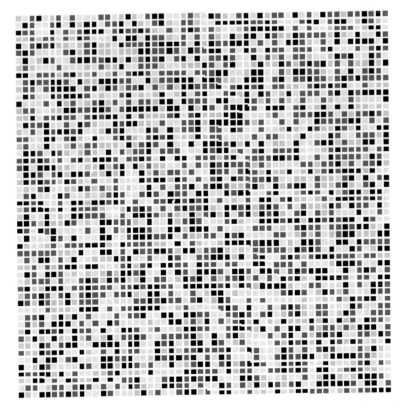

Fig. 4.14 Resulting classification using the visual cues to address the segmentation grid. By comparing Fig. 4.12 we notice that the color are properly classified after the PS channel using the proposed method with visual cues and parameters estimation with the EM algorithm

The use of proper visual cues enables the system to reach much higher payloads, at least to $2500\,\text{bytes/in}^2$, as described in this work. Moreover, it can also be combined with a non payload indicia (NPI) approach [17] for color correction and localization. It is envisioned that this hybrid technology is capable to achieve higher than $3600\,\text{bytes/in}^2$ payload density (5 by 5 pels per print code) with low bit error probability for applications that demand such a payload.

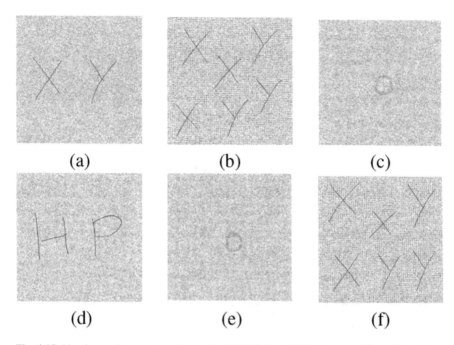

Fig. 4.15 Number and percentage of misclassified blocks (PMB) due to coffee spills and pen scribing and resulting bit errors after ECC: (**a**) 206 blocks, 0.36% PMB with BCH: 0 bits, (**b**) 556 blocks, 0.96% PMB with BCH: 9 bits, (**c**) 390 blocks, 0.68% PMB with BCH: 12 bits, (**d**) 324 blocks, 0.56% PMB with RS: 0 bits, (**e**) 341 blocks, 0.59% PMB with RS: 0 bits, (**f**) 560 blocks, 0.97% with RS: 44 bits, ©2009 IEEE. Reprinted from [9] with permission license no. 4220850864216

References

1. C. Bishop, *Pattern Recognition and Machine Learning* (Springer, New York, 2006)
2. P.V.K. Borges, J. Mayer, Text luminance modulation for hardcopy watermarking. Signal Process. **87**(7), 1754–1771 (2007)
3. P.V.K. Borges, J. Mayer, E. Izquierdo, Document image processing for paper side communications. IEEE Trans. Multimedia **10**(7), 1277–1287 (2008)
4. T.M. Cover, J.A. Thomas, *Elements of Information Theory* (Wiley-Interscience, New York, 1991)
5. N. Damera-Venkata, J. Yen, Image barcodes, in *Proceedings of the SPIE Color Imaging VIII: Processing, Hardcopy and Applications*, January 2003, vol. 5008 (SPIE, Bellingham, 2003), pp. 493–503
6. R.C. Gonzalez, R.E. Woods, *Digital Image Processing* (Prentice Hall, Upper Saddle River, 2007)
7. T. Langlotz, O. Bimber, Unsynchronized 4D barcodes, in *Advances in Visual Computing, Proceedings of the Third International Symposium, ISVC 2007, Part I*, Lake Tahoe, NV, 26–28 November 2007 (Springer, New York, 2007)
8. X. Liu, D. Doermann, H. Li, A camera-based mobile data channel: capacity and analysis, in *Proceeding of the 16th ACM International Conference on Multimedia*, October 2008, pp. 359–368

9. J. Mayer, J.C.M. Bermudez, A.P. Legg, B.F. Uchôa-Filho, D. Mukherjee, A. Said, R. Samadani, S. Simske, Design of high capacity 3D print codes aiming for robustness to the PS channel and external distortions, in *16th IEEE International Conference on Image Processing (ICIP)* (2009)

10. N. Otsu, A threshold selection method from gray-level histograms. IEEE Trans. Syst. Man Cybern. **9**, 62–66 (1979)

11. R.C. Palmer, *The Bar Code Book* (Trafford Publishing, Victoria, 2007)

12. D. Parikh, G. Jancke, Localization and segmentation of a 2d high capacity color barcode, in *IEEE Workshop on Applications of Computer Vision* (2008), pp. 1–6

13. S. Pei, G. Li, B. Wu, Codec system design for continuous color barcode symbols, in *IEEE 8th International Conference on Computer and Information Technology Workshops*, July 2008, vol. 8(11) (IEEE, New York, 2008), pp. 539–544

14. R. Samadani, D. Mukherjee, Photoplus: auxiliary information for printed images based on distributed source coding, in *Visual Communications and Image Processing*, January 2008

15. D. Shaked, Z. Baharav, A. Levy, J. Yen, N. Saw, Graphical indicia, in *Proceedings of the IEEE International Conference on Image Processing*, September, vol. 1 (IEEE, New York, 2003), pp. 485–488

16. S. Simske, J. Aronoff, M. Sturgill, Spectral pre-compensation of printed security deterrents, in *The Conference on Optical Security and Counterfeit Deterrence*, January 2008

17. S. Simske, J. Aronoff, M. Sturgill, G. Golodetz, Security printing deterrents: a comparison of thermal inkjet, dry electrophotographic and liquid electrophotographic printing. J. Imaging Sci. Technol. **52**(5), 1–7 (2008)

18. R. Villan, S. Voloshynovskiy, O. Koval, J. Vila, E. Topak, F. Deguillaume, Y. Rytsar, T. Pun, Text data-hiding for digital and printed documents: theoretical and practical considerations, in *Proceedings of the SPIE* (SPIE, Bellingham, 2006), pp. 15–19

Printed in the United States
By Bookmasters